T0250420

The Evolution of Reason

Formal logic has traditionally been conceived as bearing no special relationship to biology. Recent developments in evolutionary theory suggest, however, that the two subjects may be intimately related. In this book, William Cooper presents a carefully supported theory of rationality in which logical law is seen as an intrinsic aspect of the process of evolution. This biological perspective on logic, though at present unorthodox, suggests new evolutionary foundations for the study of human and animal reasoning.

Professor Cooper examines the formal connections between logic and evolutionary biology, noting how the logical rules are directly derivable from evolutionary principles. Laws of decision and utility theory, probabilistic induction, deduction, and mathematics are found to be natural consequences of elementary population processes. Relating logical law to evolutionary dynamics in this way gives rise to a unified evolutionary science of rationality.

The Evolution of Reason provides a significant and original contribution in evolutionary epistemology. It will be of interest to professionals and students of the philosophy of science, formal logic, evolutionary theory, and the cognitive sciences.

William S. Cooper is Professor Emeritus at the University of California, Berkeley.

CAMBRIDGE STUDIES IN PHILOSOPHY AND BIOLOGY

General Editor
Michael Ruse   *Florida State University*

Advisory Board
Michael Donoghue   *Harvard University*
Jean Gayon   *University of Paris*
Jonathan Hodge   *University of Leeds*
Jane Maienschein   *Arizona State University*
Jesús Mosterín   *Instituto de Filosofía (Spanish Research Council)*
Elliott Sober   *University of Wisconsin*

Published Titles
Alfred I. Tauber: *The Immune Self: Theory or Metaphor?*
Elliott Sober: *From a Biological Point of View*
Robert Brandon: *Concepts and Methods in Evolutionary Biology*
Peter Godfrey-Smith: *Complexity and the Function of Mind
in Nature*
William A. Rottschaefer: *The Biology and Psychology
of Moral Agency*
Sahotra Sarkar: *Genetics and Reductionism*
Jean Gayon: *Darwinism's Struggle for Survival*
Jane Maienschein and Michael Ruse (eds.): *Biology and
the Foundation of Ethics*
Jack Wilson: *Biological Individuality*
Richard Creath and Jane Maienschein (eds.): *Biology and
Epistemology*
Alexander Rosenberg: *Darwinism in Philosophy, Social Science
and Policy*
Peter Beurton, Raphael Falk, and Hans-Jörg Rheinberger (eds.):
The Concept of the Gene in Development and Evolution
David Hull: *Science and Selection*
James G. Lennox: *Aristotle's Philosophy of Biology*
Marc Ereshefsky: *The Poverty of the Linnaean Hierarchy*
Kim Sterelny: *The Evolution of Agency and Other Essays*

The Evolution of Reason

Logic as a Branch of Biology

WILLIAM S. COOPER

Professor Emeritus
University of California, Berkeley

CAMBRIDGE
UNIVERSITY PRESS

32 Avenue of the Americas, New York NY 10013-2473, USA

Cambridge University Press is part of the University of Cambridge.

It furthers the University's mission by disseminating knowledge in the pursuit of education, learning and research at the highest international levels of excellence.

www.cambridge.org
Information on this title: www.cambridge.org/9780521540254

First published 2001

A catalogue record for this publication is available from the British Library

Library of Congress Cataloguing in Publication data

Cooper, William S.
The evolution of reason : logic as a branch of biology / William S. Cooper.
p.   cm. – (Cambridge studies in philosophy and biology)
Includes bibliographical references and index.
ISBN 0-521-79196-0 (hardback)
1. Biology – Philosophy.   2. Logic.   I. Title.   II. Series.

QH331 .C8485 2001
570′.1 – dc21   00-034260

ISBN 978-0-521-79196-0 Hardback
ISBN 978-0-521-54025-4 Paperback

Contents

Foreword

This book is about how logic relates to evolutionary theory. It is a study in the biology of logic. It attempts to outline a theory of rationality in which logical law emerges as an intrinsic aspect of evolutionary biology, part of it and inseparable from it. It aspires to join the ideas of logic to evolutionary theory in such a way as to provide unified foundations for an evolutionary science of Reason.

An understanding of modern evolutionary explanation and sympathy with its aims has been assumed throughout. A prior acquaintance with the elements of symbolic logic and probability theory has been assumed as well, and some familiarity with decision theory would be desirable. Beyond that, it is my hope that philosophers of science, logicians, evolutionists, cognitive scientists, and others, will find the exposition readable.

The mathematics has been kept to a minimum. The exception is an important appendix which sets forth in mathematical detail a critical portion of the underlying formal development. My effort has been to make the theory as clear as possible, both conceptually and mathematically, with the heavier math kept separate for those who might wish to study the theory in greater depth.

The work owes much to many people. Of special note is the fact that one of the evolutionary models receiving attention (Model 5) resulted from a collaboration with Professor Robert Kaplan, now of Reed College, to whom I am deeply indebted for numerous evolutionary insights. I am grateful to Professors Ernest Adams, Bill Maron, Steven Stearns, and several referees for their valuable suggestions and criticisms of the manuscript. The book consolidates the results of earlier investigations which benefited at various stages from the comments of George Barlow, Mario Bunge, Roy Caldwell, Christopher Cherniak,

Daniel Dennett, John Endler, Baruch Fischoff, John Gillespie, Richard Griego, Paul Huizinga, Russel Lande, Richard Lewontin, John Maynard Smith, Stanley Salthe, Glenn Shafer, Dave Wake, Edward O. Wilson, Mary Wilson, and Patrick Wilson. Mention of these kind people does not imply their endorsement of what is said here. Portions of the earlier work were supported by National Science Foundation grants IST-7917566, IST-8113213, and the Miller Institute for Basic Research in Science, Berkeley.

Berkeley, January 2000
wcooper@socrates.berkeley.edu

1

The Biology of Logic

In *The Descent of Man* Charles Darwin made some remarks about 'Reason.' They begin

> Of all the faculties of the human mind, it will, I presume, be admitted that *Reason* stands at the summit. Only a few persons now dispute that animals possess some power of reasoning. Animals may constantly be seen to pause, deliberate, and resolve. It is a significant fact, that the more the habits of any particular animal are studied by a naturalist, the more he attributes to reason and the less to unlearnt instincts. ... (Darwin 1871, p. 75)

The passage continues with an astute commentary on the evolution of Reason in humans and animals.

The discussion initiated by Darwin has continued to this day. It has grown into a sophisticated discourse of considerable fascination, drawing on several disciplines. It has delved into animal reasoning in general and human rationality in particular. I have no special quarrel with the details of this extensive literature, to which I have contributed. Nevertheless, regarding the whole, I cannot help suspecting that something akin to a Ptolemaic blunder has been made. The larger order of things has been misconceived.

The original Ptolemaic blunder was rectified by the Copernican revolution, an event that has long intrigued methodologists of science. Ptolemy had the heavenly bodies orbiting a still earth. Centuries later, Copernicus changed the course of astronomy by taking the sun to be the central stillness instead. At the time there were no new observational findings to prompt the change. It was a matter of interpreting the same empirical data from a radically different standpoint. A number of subtle explanatory economies combined to support the

1

heliocentric model. The acceptance of the new theory was gradual, and was abetted by a contemporaneous questioning of Aristotelian doctrines (Kuhn 1957).

Today, in the general drift of scientific thought, *logic* is treated as though it were a central stillness. Although there is ambiguity in current attitudes, for the most part the laws of logic are still taken as fixed and absolute, much as they were for Aristotle. Contemporary theories of scientific methodology are logicocentric. Logic is seen commonly as an immutable, universal, metascientific framework for the sciences as for personal knowledge. Biological evolution is acknowledged, but is accorded only an ancillary role as a sort of biospheric police force whose duty it is to enforce the logical law among the recalcitrant. Logical obedience is rewarded and disobedience punished by natural selection, it is thought. All organisms with cognitive capacity had better comply with the universal laws of logic on pain of being selected against!

Comfortable as that mindset may be, I believe I am not alone in suspecting that it has things backward. There is a different, more biocentric, perspective to be considered. In the alternative scheme of things, logic is not the central stillness. The principles of reasoning are neither fixed, absolute, independent, nor elemental. If anything it is the evolutionary dynamic itself that is elemental. Evolution is not the law enforcer but the law giver – not so much a police force as a legislature. The laws of logic are not independent of biology but implicit in the very evolutionary processes that enforce them. The processes determine the laws.

If the latter understanding is correct, logical rules have no separate status of their own but are theoretical constructs of evolutionary biology. Logical theory ought then in some sense to be deducible entirely from biological considerations. The concept of a scientific *reduction* is helpful in expressing that thought. In the received methodological terminology the idea of interest can be articulated as the following hypothesis.

REDUCIBILITY THESIS: *Logic is reducible to evolutionary theory.*

This is intended to apply at least to the ordinary, classical theories of logic, in a standard sense of reducibility to be explained.

To paraphrase, the hypothesis is that the commonly accepted systems of logic are branches of evolutionary biology. The foundations of logical theory are biological. The principles of pure Reason, however pure an impression they may give, are in the final analysis propositions about evolutionary processes. Rules of reason evolve out of evolutionary law and nothing else. Logic is a life science. That is, of course, only an impressionistic gloss of the thesis; its exact meaning will have to be clarified as we go along.

The thesis might on first encounter seem dubious or even absurd. Certainly it is in need of interpretation and qualification. Nevertheless I hope to demonstrate that it has a core of truth that is entirely defensible. It would be too much to hope to establish it with finality in any single work, but the reasons for thinking it plausible and attention-worthy can be set forth. I beg the reader's suspension of disbelief until the chain of reasoning that supports the thesis can be laid out.

The issues involved are not vacuous. The philosophy of logic is at stake and perhaps the practice too. If as students of logic we indulge indefinitely the ancient habit of regarding logical principles as absolute and independent of biology, we will never think to look to evolutionary theory for a better understanding of them, or for ways of validating or refining them. The time may be ripe to look more seriously in that direction. If logic really is a matter of evolutionary dynamics, it should be so addressed.

It is only in recent years that it has become feasible to analyze logic from the standpoint of an advanced theory of evolution. Evolutionary biology is still young as an exact science. Parts of it have matured sufficiently by now, though, so that their ties with the foundations of logic have begun to emerge. The relationship has yet to be articulated to everyone's satisfaction, but it is sensed. This essay is my attempt to bring the ties into clearer focus, so that others may judge more easily whether a change of outlook is called for.

THE PROVENANCE OF LOGIC

Everyone will agree that something called Reason exists, is important, perhaps even "stands at the summit . . . of all the faculties of the human mind" just as Darwin said. It is also clear that this thing called Reason, whatever it may be, is based on principles called Laws of Logic. The puzzle is: *Where do the Laws of Logic come from?* That will be the topic question of our inquiry.

The answer to be proposed is that logical law comes directly from evolutionary law. That it does so is the intuitive content of the Reducibility Thesis. The hypothesis that logic is reducible to evolutionary theory is a methodologically explicit way of saying, and providing a handhold for demonstrating, that logical principles follow in the train of laws of evolution.

In case the thesis still seems obscure, the spirit of it can be illustrated with a couple of hypothetical scientific questions and answers. The first question is, "How do birds manage to fly?" A full treatise on the subject would involve two different sorts of theory. One sort would have to do with the laws of aerodynamics – the physics of gases, the viscosity of air, slipstreams, loads, lift, and so forth. The aerodynamic theory would be needed to explain how the design of the wing succeeds. The other kind of theory would concern the evolutionary considerations that brought about the flight adaptation in birds and gave it its present form. It would take up how the selective forces associated with the advantages of flight acted on genetic variation to increase fitness in the population, causing the flight adaptation to appear and be refined. Topics such as population process models, measures of fitness, and evolutionary competition would be featured in this second part. Thus the answer as a whole would involve an interplay of at least two different sorts of principles, one the laws of aerodynamics and the other the laws of evolution.

The second question is, "How do humans manage to reason?" Since the form of this question is the same as that of the first, it would be natural to attack it in a similar two-pronged fashion. One part of the answer, which might naturally be placed at the beginning of a treatise on the question, would consist of logical theory. The different kinds of logic – deductive, inductive, mathematical, etc. – would be expounded and derived from first principles, perhaps in the form of axiomatizations of the various logical calculi. These ideal systems would be taken to define the rules of correct reasoning. The explanation of how humans evolved in ways that exploit these principles would come later on. The stages of adaptation to the rules of logic would be discussed, including some consideration of how well or poorly the human mind succeeds at implementing the fundamental logical principles set forth in the first part. Somewhere in the latter part there would be talk of selective forces acting on genetic variation, of fitness, of population models, etc. As with the former question, two distinct sorts of theory appear to be involved. There would again be two parts to the exposition, a first part explaining the laws of logic and a second the laws of evolution. All this seems, on the surface at least, in good analogy with the explanation of bird flight.

What the Reducibility Thesis proposes is that it is a *false* analogy. *There are no separable laws of logic.* It is tempting to think of the power of reasoning as an adaptation to separate principles of logic, just as flying is an adaptation to separate laws of aerodynamics. The temptation should be resisted. The laws of Reason should not be addressed independently of evolutionary theory, according to the thesis. Reasoning is different from all other adaptations in that the laws of logic are aspects of the laws of adaptation themselves. Nothing extra is needed to account for logic – only a drawing out of the consequences of known principles of natural selection.

It follows that the first part of the hypothetical treatise on how humans manage to reason – the pure logic – is superfluous. The second, evolutionary, part should suffice to tell what Reason is and where the principles of reasoning come from. The prolegomenon on logic can be omitted in favor of a unified treatment in which the laws of logic emerge naturally as corollaries of the evolutionary laws.

Moreover, if this *can* be done it *should* be done. At least, it should if one believes in Ockham's razor. It is a matter of explanatory economy, which is no less important here than it was for Copernican astronomy. If the reducibility hypothesis is correct, an explanation of reasoning need not import principles of logic from some alien venue as though they were a form of knowledge peculiar unto themselves. They are already fully implicit in known evolutionary principles, waiting there to be noticed and drawn out. The laws of logic are redundant in the presence of the laws of evolution.

Because it would be easy to mistake our purpose, I had better say what the purpose is not. The aim is *not* just to show that organismic reasoning ability is a product of evolutionary forces. That much is already obvious and it is hard to see how any Darwinian could deny it. The problem with such an assertion is not that it is untrue, but that it says nothing about where the laws of logic come from. It evades the topic question. It leaves the door open to the conventional conceit according to which evolutionary pressures mold the organism to preexistent, independent, logical principles descended somehow from some rational paradise. It is the latter presumption in its various guises that I wish to oppose. According to the Reducibility Thesis there is no such rational heaven. The laws of logic are neither preexistent nor independent. They owe their very existence to evolutionary processes, their source and provenance.

THE CLASSICAL FAMILY OF LOGICS

Actually, the hypothetical treatise on how humans manage to reason is not so hypothetical. In the chapters to follow, the attempt will be made to explain organismic reasoning in a manner respectful of Ockham's razor. The aim is a unified treatment in which the laws of logic are not introduced by fiat, nor drawn from some separate philosophical foundation, but emerge inevitably from the laws of evolution themselves. Different kinds of logic will then appear as manifestations of evolutionary laws at different levels of abstraction.

Deductive logic is probably what most people first think of on hearing the word 'logic'. Deductive reasoning is the kind of logic that offers argument forms in which conclusions follow from premises with (alleged) certainty. But deductive logic, though renowned in the pantheon of rationality, is only one constituent of a greater whole. The larger logical complex involves other formalisms including general *mathematics*. Deductive logic and mathematics are so intertwined that it has seemed to many to be an arbitrary matter where one sets the dividing line between them. Looking in another direction, deductive logic is also closely tied to probabilistic or *inductive* logic, the deductive being a sort of limiting case of the inductive according to one view. Statistical reasoning then elaborates probabilistic induction. Going a step further, probabilistic logic is implicated in *decision theory*. In the theory of decision under uncertainty, sometimes also called the 'logic of decision', probability theory is enhanced by the introduction of values called 'utilities' to provide a way of reasoning about the most coherent course of action a rational agent might take.

Each of these interrelated areas of logical theory presents a facet of rationality. It is the whole complex of such systems that is referred to in the Reducibility Thesis under the cover term 'logic'. The hypothesis is that they are *all* reducible to evolutionary theory.

Attention will be confined here to the standard, or 'classical', systems of logic. They are the common theories of deduction, probability, decision, and mathematics usually presented in textbooks and elementary courses and typically applied in practice. They are the logics that most mathematicians have in mind when attempting to formalize a proof, what most statisticians regard as foundational, what consultants commonly use to analyze management decisions, what artificial intelligence researchers most often build into their programs, and so on.

There are of course other kinds of logic than the classical, but to keep the discussion within bounds they will not be considered here. The better-known nonclassical systems include intuitionistic logic, modal logic, combinatorial logic, tense logic, many-valued logic, fuzzy logic, relevance logics, and other more specialized types of formalized reasoning. Whether some of these nonstandard systems might also be reducible to evolutionary theory is an interesting and perhaps researchable question, but not one that will be addressed in these pages.

It will be seen later that evolutionary theory gives rise not only to the classical systems of logic, but also to some generalized versions of the classical calculi with nonclassical properties. The status of these unfamiliar logics will be a matter for discussion later. For the moment they are mentioned only as additional candidates for the reduction. In summary, the Reducibility Thesis as it will be taken up here asserts that all the above-mentioned classical systems of logic, and also certain associated paraclassical systems to be described, are reducible to evolutionary biology.

BEHAVIOR AS COMMON GROUND

Decision theory is the branch of logic that comes into most immediate contact with the concerns of evolutionary biology. Decision theory and evolutionary theory are bound to each other by virtue of their mutual involvement with behavior. The concern with behavioral patterns provides a common boundary region between them.

The logic of decision is concerned with an agent's choice of the most reasonable course of action from a set of available courses of action. In decision theory a course of action is called an 'act', an 'option', or in complex cases a 'strategy'. But whatever it may be called, such a course of activity is a behavioral pattern of some sort. Now, behavior is something that evolutionary theory has much to say about. Behavior is observable, it is amenable to scientific prediction and explanation, and because it is a phenotypic property of organisms the possibility arises of explaining it in evolutionary terms. This makes behavior an interdisciplinary bridge approachable from both the biological and the logical sides.

The standard systems of logic – inductive logic, deductive logic, decision logic, and so on – are so tightly interwoven that the character of the decision behavior posited in the decision-theoretic constituent of logic *determines* all of the remaining logic in the classical cluster. This may not be immediately apparent but will become clearer later. The upshot is that all of classical logic is closely tied to evolutionary theory and dependent upon it. If evolutionary considerations control the relevant aspects of decision behavior, and these determine in turn the rest of the machinery of logic, one can begin to discern the implicative chain that makes the Reducibility Thesis thinkable.

The general idea behind the reduction then is that evolutionary factors influence the character of reasoned behavior to the point of dictating it completely. Behavior is the fulcrum over which the evolutionary forces extend their leverage into the realm of logic. Viewed through the lens of biology, the behavior in question is evolutionarily fit behavior. Through the lens of logic it is rational decision behavior.

If the evolutionary control over the logic is indeed so total as to constrain it entirely, there is no need to perpetuate the fiction that the logic has a life of its own. It is tributary to the larger evolutionary mechanism. That being so, logic might as well be recognized outright as the branch of evolutionary theory that it is – momentous, but a branch nonetheless.

POPULATION PROCESSES INDUCE LOGICS

By *biology* we shall usually mean evolutionary biology. Within evolutionary biology the narrower focus will be on *population biology*, widely considered to be the mathematical core of evolutionary biology. Population biology includes the formal study of population process models, population genetics, selection, adaptation, and evolutionary fitness. The reducibility hypothesis could have been reworded to assert that logic is reducible to population biology.

The interplay between logic and biology comes down to this. Theories of population biology, when made precise, take the form of mathematical population process models and the properties deducible from them. One of their deducible properties, it will be seen, is that they spawn rules of logic. That is, particular population theories entail not just a tendency on the part of fit population members to obey external logical constraints, but the logical rules themselves. The population models determine what fit behavior shall be, under the conditions postulated; and this fit behavior, regarded as decision behavior, determines the logic.

In this way the general evolutionary tendency to optimize fitness turns out to imply, in and of itself, a tendency for organisms to be rational. Once this is shown there is no need to look for the source of logical principles elsewhere, for the logical behavior is shaped directly by the evolutionary forces acting on their own behalf. Because the biological processes expressed in the population models wholly entail the logical rules, and are sufficient to predict and explain rational behavior, no separate account of logic is needed.

DEFINING REDUCTION

To say that logic is a 'branch' of biology, or that biology and logic are candidates for a 'unification', or that population processes 'induce' systems of logic, and so forth, is to speak loosely of a relationship that can be described more precisely as a *reduction*. Reduction has an honored place in science. It has been described as "the explanation or replacement of one scientific theory or branch of science by another" (Schaffner 1977, 146). It involves the grafting of one theory onto another in such a manner that the composite result is more economical of concepts and laws than the sum of the two original theories.

Historic examples of reducibility relationships include Newton's reduction of Kepler's planetary equations to the general laws of motion and universal gravitation, the reduction of Galilean mechanics to the same, the reduction of thermodynamics to statistical mechanics, and the reduction of parts of chemistry to particle physics. Mendelian genetics, or extensions of it, are thought by some to be largely reducible to molecular genetics. Methodologists still debate the details of these famous reductions, but few doubt that they are indeed scientific reductions in some sense or to some extent. In their time, all were first-class "Aha!" experiences.

The kind of reduction that will be relevant here is epistemological or nomological reducibility, or what is sometimes called *theory-reduction*. The general idea of a theory-reduction is that one theory is reducible to another just in case it can be derived from it by logical or mathematical steps without introduction of fresh subject matter. The simplest explicit characterization of theory-reducibility is the well-known model due to Nagel (1961) and others. Omitting some refinements, a scientific theory T_2 is said in the Nagel model to be *reducible* to another theory T_1 if and only if (1) the concepts of T_2 can be defined in terms of concepts of T_1, and (2) using these definitions, the propositions of T_2 can be deduced from the propositions of T_1. Clauses (1) and (2) are the so-called conditions of *connectability* and *derivability*.

It is understood that the two operations can proceed in any number of stages. New concepts can be defined within the reducing theory, then new propositions can be derived with the help of the concepts just defined, then more new concepts can be defined with the help of the newly derived theory, and so forth until the theory to be reduced is eventually arrived at. Theory-reduction is a matter of derivability in any number of stages with allowance made for creativity of definition.

The formal Nagel definition is an oversimplification of what many actual theory-reductions are like. Some believe it to be a gross, even hopeless, oversimplification (Burian 1985, 25; Schaffner 1977). But it is adequate for simple cases and conveys the spirit of more involved reductions. Generalizations of it have been proposed and they may indeed appropriately broaden the scope of the original (Schaffner 1993). However, as it applies to the Reducibility Thesis, the modifications and extensions to the Nagel model that are needed are probably minimal. It would be premature to go into the exact formal details of the kind of theory-reduction required. Nagel's original characterization comes close enough to what is wanted for purposes of preliminary exploration.

The reduction of interest should not be confused with another kind of reduction to which it bears only a distant resemblance. There is a commonplace form of reductionism encountered in biology and ecology in which the properties of a biological system are analyzed hierarchically in terms of the properties of the system's members, physical components, or ecological subsystems. It is the type of reducibility referred to by G. C. Williams (1985) when he wrote "Reductionism is the seeking of explanations for complex systems entirely in what is known of their component parts and processes." The reduction of concern here is not of this mechanical kind. Although it may be possible to cast the sorts of reductions Williams refers to in the Nagel form, not all Nagel reductions are based on physical componential analysis. The theory-reduction of present interest has little to do with physical part–whole relationships and much to do with derivation and definition.

10

In comparing biology with logic, one has population process models on the one hand and systems of logic on the other. Both can (ideally) be cast in exact mathematical terms. The nub of the reducibility claim is that from each population model regarded as a formal biological theory one can derive, using the mathematics of fitness optimization, an associated theory of logic. In so doing, one establishes logical principles on the basis of evolutionary precepts. The logical principles so derived are local to the particular population model, but – or so it will be argued – no less logical for that.

We shall not be able to establish the reducibility claim definitively here by producing complete derivations of logical systems from population models in full mathematical detail. The Nagel model assumes that in principle the reduced and reducing theories can be formalized explicitly and completely, ideally in axiomatic form, and that the reduction supplies a complete formal deduction of the reduced theory from the reducing with every tiniest deductive step in order. However, Nagel recognized and explicitly stated that this is an "ideal demand" not usually satisfied in the normal course of scientific reductions. More often a reduction is largely conceptual with exact derivations given only for certain critical portions of the chain of reductive reasoning. That is the precedent to be followed here. Formal definitions will be offered and relevant theorems stated that are proved either here or by other authors. With some informal reasoning to connect them, these will come together to constitute an argument for reducibility.

Reductionism has had its critics, especially in biology; and the critics are able to point to abuses. The criticisms have to be taken seriously. It should be remembered, though, that it is the abuses that are blameworthy, not the reductive method itself.

REDUCTIONISM IN LOGIC

In the common Darwinian view, all human and animal capabilities, including mental capacities, are biologically evolved. Logical reductionism agrees but goes beyond that obvious point. The additional claim it makes is that there is a direct dependency of the laws of logic on the laws of evolution – a sort of homomorphism from evolutionary theory to logical theory. The evolutionary laws extend across the boundary of behavior to control the logic.

If the claims of evolutionary reductionism can be sustained, logical laws are not just products of historic evolutionary processes, but are themselves formulable as part of the theory of those processes. Not only do laws of logic evolve, they are also partial descriptions of what it means to evolve. A logical schema becomes a kind of evolutionary equation or proposition, albeit heavily disguised.

If this be the case, evolutionary biology is unique among the sciences as the seed bed for the laws of logic. If the various sciences are likened to factories of knowledge – factories that use logical tools – then population biology is the tool factory. Biology is an ordinary science but its evolutionary branch is a subscience that, in addition to its more ordinary offices, predicts and explains logic.

According to the reductionist claim, logic is so biological that if the classical laws of logic had not already been worked out independently, an evolutionist innocent of any prior knowledge of formal logic could in principle have stumbled upon them simply by drawing out the consequences of standard evolutionary models and processes. Starting in the next chapter we will put ourselves in the position of such a logically naive evolutionist in order to witness the extent to which the drawing-out can actually be accomplished.

LOGIC AS SCIENTIFIC GENERALIZATION

In the usual view, the laws of logic are independent truths to which organisms must adapt. Complex animals are considered to have evolved in such a way as to implement to some extent these preexisting rules of reason. The more cognitively advanced the organism, the better the implementation, it is thought.

This 'implementation' is considered to take place through the evolution of neural mechanisms or other physiological means, but in any case, the process of adaptation to the rules has customarily been regarded as something distinct from the rules themselves. In maintaining that distinction, Nature is cast as an engineer who designs a computer to implement independently valid logical and arithmetic truths. The engineer designs the computer but not the truths. It is conceded that Nature does her designing differently from a human engineer, by an eons-long process of genetic trial and error in fact, but still the logical constraints are considered to be already set forth independently of anything Nature-as-designer may do. Evolution has to accommodate itself to the logical boundary conditions.

The reductionist way of thinking calls this paradigm into question. In the reductionist view the evolution of rationality is not at all a matter of organismic brains 'implementing' preexisting logical rules or 'adapting to' them. Instead, it is a matter of rules of logic coming into existence as concomitants of the adaptive process itself. Nature may be an engineer, but she is not the kind of engineer who works to implement preconceived design goals. Rather, this engineer engineers blindly and madly, without prior task specifications, and whatever gets engineered gets engineered. If rational organisms evolve, then the properties that prompt us to call them rational must have come out of the engineering procedure itself. They could hardly have come out of independent design requirements that Nature felt obliged to impose out of mysterious metaphysical loyalties.

Older tradition has it that rules of logic have a sovereign, transcendental validity that is in no way dependent upon any particular empirical science. Admitting that it is generally fit to reason correctly, the conventional view concludes that evolutionary pressures cause the behavior of sufficiently complex organisms to manifest an approximation to these rules. What is wrong with this, from the reductionist perspective, is the subtle assumption that the evolving ratiocinatory powers conform to logical constraints that are *extrabiological*. This unwarranted notion has falsely endowed logic with its own private criteria of validity, as though it were a law unto itself.

In the reductionist perspective, logic is not extrabiological but wholly emergent from evolutionary processes. There is no crisp category distinction to be made between logical and biological truths. There is no such thing as 'pure logic', if what is meant by that is a logic independent of empirical facts and physical processes. To the contrary, all logic is thoroughly impure and biologically contaminated from the outset.

HINTS FROM THE LITERATURE

An evolutionary outlook on logic and rationality has been adopted by many writers, and many of the ideas to be dealt with here have been around a good while. As far as I am aware, I am the first author who has been rash enough to have suggested in print that logic is reducible to evolutionary theory (Cooper 1987, 1988). Others have come to the brink however. Intimations of the biological character of logic are to be found scattered throughout the literature of several disciplines. Though a comprehensive survey of individual works would be too vast to contemplate, a few areas of inquiry can be pointed out as especially germane.

First there is the literature on evolutionary biology itself. Especially relevant are the parts that involve game theory and decision theory. Outside of biology, game theory is intended to formalize what is involved in acting reasonably in interactive choice situations, the idea being to capture an aspect of rationality in mathematical terms. Within biology (e.g., Maynard Smith 1982; Maynard Smith and Price 1973) game theory is given the additional twist that it is used to describe something that evolves. In the evolutionary context it is clear that game theory is not purely *a priori* but also empirical and biological, for the precise form of a game-theoretic logic is clearly dependent upon the relevant biological circumstances. (For surveys see e.g., Maynard Smith [1978, 1982] and Reichert and Hammerstein [1983]).

Decision theory is an elementary form of game theory. Works on game theory frequently present decision logic as a special case of game theory, namely, the case in which the player's 'opponent' happens to be a neutral Nature. Decision- and game-theoretic vocabulary was introduced into the evolutionary literature by Lewontin (1961). Explicit correspondences between decision-theoretic and evolutionary concepts have been listed and discussed, for example, by Templeton and Rothman (1974). When the decision-theoretic elements of evolutionary analysis were analyzed in greater detail and compared with those of classical decision theory, the idea of a 'natural decision theory' emerged (Cooper 1981). Natural decision theory was conceived as the evolutionary analog of standard decision theory, the difference being that in natural decision theory the role of decision maker is played by natural selection. Later contributions have used decision-theoretic formalisms in the evolutionary context in various ways. A paper provocatively entitled "Darwin Meets the Logic of Decision" (Skyrms 1994) and other works by the same author analyze the formal analogies – along with some disanalogies – between evolutionary dynamics and standard theories of decision and games (Skyrms 1996, 1997).

Decision theory is intimately related to probability theory. A famous treatment blending the two is contained in a book by Savage (1972), a remarkable work that extracts subjective probability theory from an austere behavioral basis. Savage's system is an embodiment of the important idea that probabilistic induction rests on decision- and utility-theoretic foundations. In present terms it is essentially reducible to them. There is, as yet, no significant literature on evolutionary probability theory as there is on evolutionary game and decision theory, but such a subject suggests itself because of the reductive connections between decision theory and inductive logic worked out by Savage and others.

Probability theory, the formal basis of statistical inference, is in turn intimately related to deductive logic. The connection has been formalized in various ways. One account has it that the classical deductive logic is the special case of inductive logic in which attention has been confined to inferences in which the conclusion may be reached with a probability approaching certainty. More elaborate analyses of the relationship include the one provided by Carnap and Jeffrey (1971).

According to a philosophical school known as Logicism founded by Gottlob Frege, Bertrand Russell, and Alfred North Whitehead, general mathematics is reducible to deductive logic. Much has been written both for and against this position. The discussion becomes relevant here if logic can itself be shown to have biological foundations. If it does and if logicism is accepted, then mathematics must have biological foundations too.

Still other disciplines bear on the reducibility thesis. In the economics literature, many a study is inhabited by an idealized creature called 'rational man' or 'economic man'. The penchant for analyzing rational economic behavior has led to fundamental contributions to logic, especially the already-mentioned logic of decision and utility theory and game theory, the latter having received much of its initial impetus in economics (von Neumann and Morgenstern 1953). Economic papers have occasionally appeared that interpret the notion of economic or decision-theoretic utility explicitly in terms of evolutionary fitness (e.g., Robson 1996). Pockets of the economic literature on decision and game theory resonate with evolutionary overtones for those so attuned.

There is a voluminous psychological literature about logic and decision making as actually practiced by humans and animals. Points of contact with evolutionary theory are frequently discussed. The literature contains a body of experimental work comparing actual human and animal reasoning with the reasoning predicted by classical and other systems of logic. There are findings of systematic divergences

15

between classical and actual reasoning and these have been the subject of much speculation. Doubtless many of the observed discrepancies signify nothing more than slips or blunders, but in some cases the experimental subjects are not only consistent with one another in their deviant responses but stoutly refuse to change their responses even after having had the "correct" classical reasoning explained to them.

In such circumstances the possibility exists that the subjects are using a biologically valid system of logic that simply happens to differ from classical norms. These anomalies are grist for the mill of biological reducibility. Under the hypothesis of reducibility, some of the non-classical elements of the systems of logic used by humans could well be explainable on the basis of more realistic population process models than the ones that give rise to the classical logic. The observed non-classical reasoning might sometimes be predictably fitter, evolutionarily, than standard logical reasoning.

Philosophers of biology have provided thought-provoking discussions of the phenomenon of rationality, and their discussions constitute the setting for the present study (e.g., Sober 1981). They concern an aspect of what has come to be known as *evolutionary epistemology*, the philosophy of knowledge considered in the light of evolutionary theory. The paper by Campbell (1974) has been seminal, and Ruse (1986a, 1989) comments directly on the evolutionary epistemology of organismic reasoning. Much of what has been said by contributors to this literature reinforce the viewpoint that reasonableness is relative to evolutionary circumstances.

Konrad Lorenz, considered a founding figure of evolutionary epistemology, wrote (1941):

> Kant's statement that the laws of pure reason have absolute validity, nay, that every imaginable rational being, even if it were an angel, must obey the same laws of thought, appears as an anthropocentric presumption.
> . . . Nothing that our brain can think has absolute, *a priori* validity in the true sense of the word, not even mathematics with all its laws.

Such denials of an absolute *a priori*, and explicit recognition of the biological relativity of logic and even mathematics, presage reductionist systems such as the one to be developed here. Authors such as Lorenz have anticipated our theme, though it remains to be shown in reductionist terms just why they are right.

The relationships among these various theories have been the subject of investigations by evolutionists, statisticians, economists, logicians, metamathematicians, and philosophers of the highest competency and reputation. In some cases the connections have been formalized rigorously, with results proved as theorems to the most exacting standards. The chief obstacle to comprehending the evolutionary character of logic therefore lies not so much in the lack of foundational work on any given discipline, nor even in the various interdisciplinary links taken pairwise, but rather in the general fragmentation of the different studies. Each has been carried out in isolation using a local vocabulary, creating the illusion of bounded disciplinary regions. But if the time has arrived to contemplate a broader unification, it is at least encouraging that the pieces of the puzzle have already been examined separately with great care. It remains only to fit them together.

SUMMARY

From antiquity, various philosophies of logic have been proposed to explain the origin and character of rational thought. Some have given rise to elegant formal symbolic systems that allegedly codify precise logical principles, most prominently the various classical logical calculi. The foundations of the systems have been laid on motivating ideas ranging from faith in a rational intuition to theories of truth and semantics. Without necessarily discarding any of these considerations as irrelevant, we have suggested that there may be a more comprehensive approach to the foundations of logic in which logic is developed as a subscience of evolutionary theory.

Such a development is feasible if it can be shown that the principles of logic can be derived directly from evolutionary propositions. That this is possible is a hypothesis called here the Reducibility Thesis. It states that the laws of logic, or at least of classical logic and certain generalizations of it, are reducible to evolutionary biology in a standard sense: The terms of the logical theory are definable in evolutionary terms and logical assertions are deducible from evolutionary assertions.

If the Reducibility Thesis has merit, the principles of rationality are so deeply embedded in evolutionary theory that their foundations cannot rigorously be investigated independently of it. The motive for seeking these deeper foundations is not just that the capacity to reason is produced by evolutionary processes, a fact now well accepted by the modern mind. It is that the underlying rules of reasoning are themselves recodifications of the properties of the processes.

The reductionist claim will be regarded with deep skepticism by many. It has long been customary to set forth systems of logic as though they were independent of any particular empirical science. Traditionalists might for that reason think it perverse to bestow on evolutionary science any special formative status with respect to logic. Indeed, to scholars unfamiliar with the associations between logic and evolutionary theory, the very thought of logic as a department of biology must seem bizarre. But those who are inclined to reject the idea out of hand should be asked, before passing judgment, to examine the chain of biological steps leading from the evolutionary premises to the logical theory.

2

The Evolutionary Derivation of Life-History Strategy Theory

The thrust of the Reducibility Thesis is that upon pursuing the implications of evolutionary principles far enough one arrives at laws of logic. The evolutionary considerations are asserted to give rise to the logical rules themselves, not merely to a tendency for organisms to obey externally imposed logical rules of mysterious origin. To demonstrate the thesis the logical theory must be extracted directly from the evolutionary theory. In so doing, care must be taken that the logic does not somehow get smuggled in the back door. It must be clear that the logic can be developed as an inherent part of evolutionary theory until the logic stands complete as an entity of biological origin.

The task of this chapter is to prepare the ground for that enterprise by reconstructing a portion of evolutionary theory to the point where it can serve as a platform on which to erect logical systems. But first let us look ahead at where this line of exploration will lead.

THE LADDER OF REDUCIBILITY

Can one start out with evolutionary theory and, by carefully drawing out its consequences, end up deriving systems of logic? We will try to show that this can indeed be accomplished in a chain of reductive reasoning involving several stages of ascent. An attempt has been made to make the stages correspond more or less to familiar bodies of theory in biology and logic.

Figure 2.1 shows the ladder of reductive relationships. The upward-pointing arrows on the left indicate implicative relationships. These run

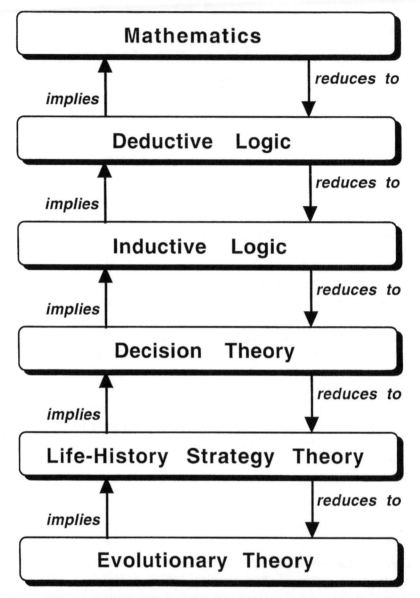

Figure 2.1. The ladder of reducibility. Each level theory-implies the one above and is reducible to the one below.

from bottom to top. Thus the ladder asserts that from general evolutionary theory one can derive a special branch of population biology known as life-history strategy theory. It in turn implies decision theory, which in turn implies inductive logic or probability theory, and so on up through deductive logic and finally general mathematics. The downward-pointing arrows on the right are the corresponding reducibility relationships. Starting at the top, mathematics is represented as reducible to deductive logic, deductive to inductive logic, and so on downward.

The term 'implies' beside the upward arrows means *theory-implies*. It is intended in the sense of the Nagel model or one of its generalizations, in which one theory implies another if the terms of the second can be defined within the first and its propositions derived from the first. With this understanding, theory A reduces to theory B if and only if B theory-implies A. Thus the upward-pointing arrows are redundant in the presence of the downward arrows and vice versa.

Lest there be unrealistic expectations, it should be added that it will be possible to present here only the most minimal treatment of the various rungs. At the decision theory rung, not every topic that ever comprised a chapter in a textbook on decision science will be explored, but only the classical logic of decision under uncertainty. It is usually the centerpiece of such expositions. Game-theoretic extensions of it will not be considered. Similar remarks apply to inductive logic. Not every system ever described as 'induction' will be examined, but only probability theory and probability-theoretic systems of induction. Nor will encyclopedic comprehensiveness be attempted for deductive logic or mathematics.

It can be said though that the systems to be examined are in some sense the critical ones. They form the central thread of the serious intellectual tradition of how rationality ought to be codified. If a few prominent calculi had to be chosen as representative of respectable logical practice, those singled out for consideration here would be among the obvious candidates. Though our ladder will be only skeletal, it is a skeleton of moment on which much else hangs.

The plan is to climb the ladder at the rate of one rung per chapter in a bottom-up reconstruction of this skeletal logic. The present chapter will take the first upward step. It systematizes the relevant portion of general evolutionary theory (bottom rung), and deduces from it an appropriate version of life-history strategy theory (second rung).

MODEL 1: CONSTANT GROWTH

Population biology studies the growth and changing makeup of evolving populations of organisms. In population biology a particular combination of assumed evolutionary conditions is represented by a *population model* – a formal mathematical representation of the circumstances that govern evolutionary changes in a population of interest. Mathematical population biology or population genetics is largely the study of such population process models and their properties. A primitive population process model called *Model 1* will serve as the basis for the initial discussion.

Model 1 is a constant growth model. The rate of increase in population size is treated as though it were a fixed proportional increase per unit of time. This is a strong simplifying assumption. As Darwin pointed out, any real population increasing unchecked at such a fixed rate (essentially exponentially) would presently overflow the planet. However, in some actual circumstances populations do increase at an approximately fixed rate for many generations, so models based on the constant growth assumption are not wholly devoid of practical application. More importantly, a constant growth model can serve as a convenient starting point from which to construct other more realistic models.

Model 1 also assumes lock-step seasonal breeding. A regular seasonal or periodic life cycle (e.g., the year) is posited, with all offspring produced at the same point in this cycle (e.g., in the spring). The constant rate of population growth can then be defined as the multiplicative factor R by which the population increases in each season. That is, R is the ratio of the population size at the end of each seasonal cycle to the size at the end of the previous cycle. R is called the *finite rate of increase* (commonly denoted R_0 or λ).

With R so defined, the basic modeling equation for any Model 1 population or subpopulation becomes

$$N(t+1) = RN(t)$$

where $N(t)$ denotes the population size at the end of the tth season. The growth of such a population follows an approximately exponential curve. It is not strictly exponential because in a fine-grained view it would be seen to grow in seasonal jerks. (Due to the jerks the finite rate of increase of discrete models such as this is not quite the same thing as the instantaneous rate of increase or 'Malthusian parameter' for continuous growth models. The latter is defined as the derivative $r = dN/dt$. See e.g., Lomnicki [1988].)

Model 1 also assumes *nonoverlapping generations*. Each population member produces offspring at most once per lifetime, at the appointed time in the season (e.g., in spring). This is called *semelparous* reproduction, as opposed to *iteroparous* reproduction in which an individual can produce offspring more than once.

It will also simplify things to assume for the present *asexual reproduction*, as in a uniparental cloning process. Asexual reproduction can actually take place for many generations at a time in some species. For most species likely to be of interest, though, it is merely a convenient simplifying assumption for a starting model.

It is easy to see that under these assumptions the growth rate R is equal to the average number of surviving offspring per parent. If, for example, the seasons are defined as beginning and ending just before reproduction time, R is the average number of offspring surviving to reproductive age.

Finally, a Model 1 population is assumed to be large. The assumption of a sufficiently sizable population allows certain complications of statistical sampling to be ignored. The vagaries of chance may affect different population members differently, but in the aggregate the rate of population growth can be taken as approximately constant because of the large population size.

Model 1 corresponds more or less to the simplest textbook population models (e.g., Crow and Kimura 1970, 5). Meager as it is, it will serve as the reference model for the next several chapters. For the sake of the exposition we will pretend that animals with complex cognitive capacities could evolve within the conditions described by the model.

TREE DIAGRAMS

Usage of the terms *character* and *trait* has varied, but here they will be used interchangeably to refer to any phenotypic properties of individuals that may be of interest, whether morphological, physiological, or behavioral. It is assumed in Model 1 that all characters or traits singled out for analysis are heritable in the strict sense that if a parent possesses the character, all its nonmutant offspring will also possess it. The inheritance mechanism is presumably genetic but need not necessarily be so. Behavioral characters, for example, might be inherited through parental training.

Suppose each member of a population has one of two possible characters, and that the two disjoint subpopulations so defined have different constant finite rates of increase. Which trait is advantaged? Obviously the subpopulation with the higher constant growth rate will take over larger and larger proportions of the total population, eventually overwhelming the remainder. It is clear then that in Model 1 all reasoning about the relative long-term success of traits boils down to calculating and comparing their respective finite rates of increase.

As an example of such a calculation, suppose the question has arisen whether it would be evolutionarily advantageous for a certain small desert animal to have a hard shell. Imagine that at a certain point in its life cycle, before it can reproduce, the organism runs the risk of encountering a certain type of predator. The virtue of a shell is that it offers some immediate protection against the predator in the event the animal is discovered. The drawback of a shell is that it encumbers the animal generally, making all locomotion more awkward. Slower movement also means the individual is exposed longer at critical moments when predation is possible, making it likelier that it will be detected by a predator in the first place.

To illustrate the possible tradeoffs, suppose that for a shelled individual the probability of being discovered by a predator is 0.4, whereas without a shell greater mobility reduces this probability to 0.3. For the subset of population members discovered by a predator, let us assume a subpopulational rate of increase of $R = 0.9$ in the case of individuals protected by a shell, and $R = 0.2$ for those not so protected. For individuals who do not encounter a predator, let the growth rate be $R = 1.4$ if they are encumbered by a shell, and $R = 1.5$ if not. Is a shell advantageous?

The data can be displayed in the form of a tree diagram of the kind shown in Figure 2.2a. (Diagrammatic conventions are from Cooper [1981].) At the root of the tree is a square node out of which two paths emerge. The paths represent the two traits to be compared (SHELL versus NO SHELL). Since the question is which trait will be selected for, the square node might be called the *selection node*. The round nodes represent environmental variables, in this instance the binary variable of whether or not the environment presents the individual with a predator encounter. The paths emerging from each round node are labeled with the possible values of this environmental variable

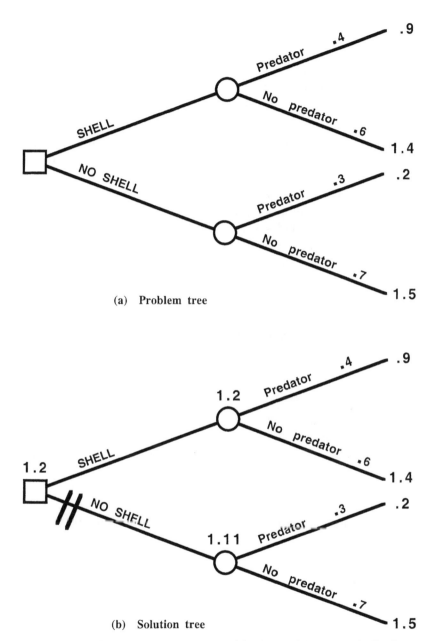

(a)   Problem tree

(b)   Solution tree

Figure 2.2. A life-history strategy tree. (a) The problem: Is a shell advantageous? (b) The solution: Yes.

(Predator or No predator) together with their probabilities of occurrence. At the four branchtips of the tree are the subpopulational growth rates (values of *R*) experienced by the four classes of individuals so differentiated.

Implicit in the diagram is an assumption that the probabilities of occurrence associated with the values of an environmental variable remain constant from season to season and independent from individual to individual. This is a statistical simplifying assumption that may be regarded as an additional defining condition for Model 1. It is as though for each round node there were a chance device such as a coin or die that is flipped or rolled separately and independently for each individual. In this example Nature determines predator encounters by flipping a coin weighted 4:6 for each shelled individual, and a coin weighted 3:7 for each unshelled individual.

The resolution of the problem is shown in the 'solution tree' of Figure 2.2b. It is constructed by performing the following computation in the 'problem tree' of Figure 2.2a: At each round node, add up the connected branchtip values to the right after weighting them by the probabilities found along the connecting paths, and write this weighted sum over the node. Thus, for example, the upper round node is assigned the value

$$R_{shell} = 0.4 \times 0.9 + 0.6 \times 1.4 = 1.2$$

which is the rate of increase of the shelled population. The reader will readily verify that in Model 1 this probability-weighted averaging procedure produces correct growth rates for the two trait-defined subpopulations.

Since the upper round node has the larger number, the conclusion to be drawn from the analysis is that shells are favored by natural selection. This number is written again over the root node to indicate that if the selective forces have their way, the entire population will end up with that rate of increase. A double bar is drawn across the lower branch to indicate that the competing character, lack of a shell, is selected against. Shell-lessness is an evolutionary path that is 'blocked off' in the presence of shelled competition.

In this example the tree is binary, but in general both selection nodes and environmental variables can be *n*-ary. Among the two or more mutually exclusive possible traits emerging from the selection node, barring ties all but the one with the maximum growth rate is blocked off, i.e., selected against.

POPULATION FLOW

The probability figures along the paths emerging from a round node can be interpreted as approximate subpopulational proportions. For instance, in Figure 2.2 each shelled individual has probability 0.4 of encountering a predator. Hence approximately 40% of the shell-equipped population members can be expected to encounter a predator. This follows from the Model 1 assumptions that the population is large and that the tree probabilities are constant and independent from individual to individual. If a coin weighted to give 0.4 probability of heads is flipped many times, it is to be expected that about 40% of the tosses will come up heads.

This suggests a way of interpreting the tree diagrams. One can think in terms of a *populational flow* of individuals coursing through the tree from left to right in each generation as evolutionary time goes by. The flow divides at all nodes. At selection (square) nodes the flow is sorted by trait. In the example, the sorting at the root node causes shelled members to flow up the upper path and unshelled down the lower. At round nodes the population is sorted by the events experienced. Thus of the (shelled) population flowing along the upper path out of the root approximately 40% encounter a predator and take the upper exit out of the round node while the remaining 60% take the lower exit path. Flows reaching the branchtips consist of subpopulations defined by the traits and environmental variable values the flow has passed through along the way. The *R*-values at the branchtips are the finite growth rates of these subpopulations.

The flow is cyclical in the sense that each generation's offspring loop back to constitute the population that flows through the tree in the next generation. As evolution proceeds the flow grows disproportionately heavier through the trait paths favored by natural selection. Eventually the selective forces, if unopposed, have their way and the advantaged traits take over.

As an illustrative computation of populational flow, suppose a hypothetical population governed by the process of Figure 2.2 is initially half shelled and half unshelled, i.e., there is a starting ratio of 1:1. The population enters the tree from the left and is sorted by the root selection node into its shelled and unshelled subpopulations. When the shelled half reaches the upper round node, approximately 40% flow up to the branchtip where their numbers are multiplied by 0.9, while the remaining 60% flow down where they are multiplied by 1.4. Thus the shelled subpopulation becomes $0.4 \times 0.9 + 0.6 \times 1.4 = 1.2$ times as large as it was at the start of the season. By a similar computation the unshelled subpopulation grows by a factor of 1.11. The total population, with proportions now weighted 1.2:1.11 in favor of shells, cycles back and goes through the tree again next generation to become still more shell-weighted, this time by a ratio of $(1.2)^2:(1.11)^2$. The generation after that the ratio becomes $(1.2)^3:(1.11)^3$, and so on as the shells gradually overwhelm the competition.

CHARACTER COMBINATIONS

The tree-diagram method is easily extended to apply to combinations of traits and environmental variables in any numbers. As a simple example, suppose it is not only shells about which one is curious, but also the question whether it is more advantageous for individuals to defend themselves from an approaching predator by burrowing quickly down into the sand or by fleeing over its surface. The two issues could be related, for it is possible that whether digging or fleeing is preferable depends on whether the organism has a shell or not. The morphological and behavioral problems must be considered together and there are tradeoffs to consider.

Figure 2.3, an extension of Figure 2.2, presents some hypothetical data by way of illustration. The choice of whether to dig or flee is represented as a new square selection node appearing in the paths of both shelled and unshelled subpopulations that encounter a predator. No matter which instinct is present, DIG or FLEE, the individual will either be 'Consumed' by the predator with a certain probability, or 'Escape' capture with the complementary probability. These values are represented by paths emerging from new round nodes for a second environmental variable.

As before there is an R-value at each branchtip indicating the rate of increase of the subpopulation that flows to it. The rate of increase for a 'Consumed' subpopulation is zero because all its members are predated before they are able to reproduce. Subpopulations that 'Escape' have the same rate of increase as those of similar shelled or unshelled character that do not encounter a predator.

Figure 2.3 is the combined problem/solution tree for the example. Values of R have been computed for all the nodes by a procedure that

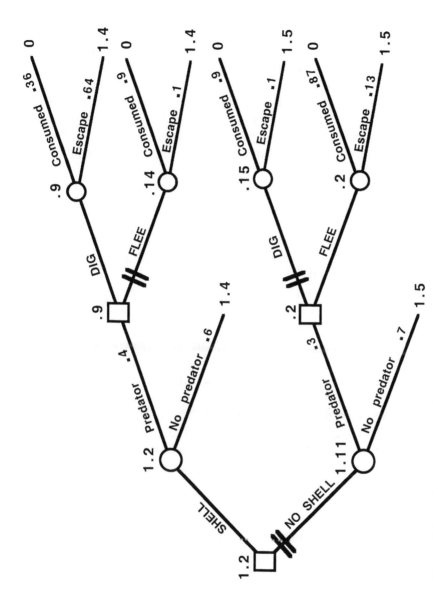

Figure 2.3. A life-history strategy tree for determining which of a combination of traits is selectively advantaged.

29

can now be stated in full generality as it would apply to any such tree however complex. Starting at the branchtips and working from right to left through the tree, perform the following operation at each node: (1) if the node is round, calculate for it the probability-weighted average of all values over connected neighboring nodes or branchtips to its immediate right, the weights being the probabilities appearing along the paths leading to them, and write the result over the node; (2) if the node is square, find the largest of the values over connected neighboring nodes or branchtips to its immediate right and write this over the node; and (3) when done, draw a double line across all the paths that lead rightward out of a square node to neighboring nodes or branchtips with nonmaximal values.

One concludes from Figure 2.3 that the trait set SHELL + DIG is selected for and all other combinations are selected against. For this winning combination the finite rate of increase is 1.2. The latter figure is displayed over the root node, for if natural selection takes its course the whole population will eventually approach this rate of increase as the shelled-and-digging form takes over.

Within the confines of Model 1, the tree-analysis method is powerful and convenient. Any number of characters can be considered in any combination, along with any number of environmental variables that might influence their success. Nodes can be binary, *n*-ary, or even (with some diagrammatic awkwardness) continuous. The root node can be either square or round, and square or round nodes can appear in any order along the branches as may be called for by the problem situation. The paths emerging rightward out of a round node must always represent mutually exclusive and exhaustive possibilities for the values of the environmental variable involved, with probabilities summing to one. Beyond that there is no restriction on the kinds of environmental conditions that can be considered so long as they conform to the statistical constraints of the model. The characters analyzed in a tree can consist of any mixture of structural, morphological, physiological, behavioral, or other heritable phenotypic traits.

CONSEQUENCES

Different combinations of traits and environmental circumstances can give rise to different *consequences*. For instance, a trait for under-

ground nest-building subjected to an environmental condition of tor-
rential rain might have the consequence DROWNED; a certain brows-
ing strategy combined with plentiful food and fine weather could have
the consequence HEALTHY; and so forth. Although it is not essential
to the method, the significance of a tree diagram can be clarified by
including at the end of each branchtip such a consequence – the end
result experienced by every member of the subpopulation that flows
to that branchtip.

In Figure 2.4 the consequences in Figure 2.3 have been made
explicit. Population members that are discovered by a predator, and
are subsequently consumed, perish before they can produce offspring.
The consequence PREDATED therefore appears at these branchtips.
Members with shells who are undiscovered by any predator, or who
having encountered a predator escape predation, survive the predator
stage but lead the rest of their lives encumbered by a shell, a consequence indicated in the diagram by the label ENCUMBERED.
Those lucky individuals without shells who are undiscovered or
escape predation anyway are said to experience the consequence
UNENCUMBERED.

As before, so also in the consequence-equipped tree, R-values
appear at the branchtips. However, in a tree equipped with consequence labels, such a branchtip value can also be conceived as the
growth rate associated with the accompanying consequence. Thus there
is a valuation function over the set of all consequences. Such a function assigns a unique growth rate to each consequence in the set. In
the example of Figure 2.4 the consequence set is the three-member set
PREDATED, ENCUMBERED, and UNENCUMBERED. The valu-
ation function on it is as follows.

Consequence	Value of R
UNENCUMBERED	1.5
ENCUMBERED	1.4
PREDATED	0.0

In general, a consequence can be any outcome or final state that
rewards, penalizes, or otherwise characterizes the experience of a
branchtip subpopulation. Since it need only be specified narrowly
enough to determine the associated growth rate, the same consequence
can appear at more than one branchtip.

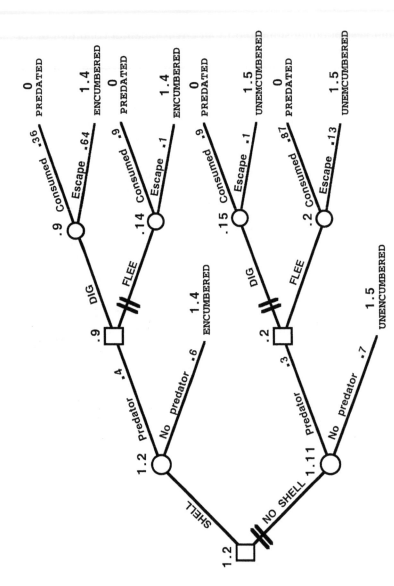

Figure 2.4. The tree of Figure 2.3 with branchtip consequences labeled.

32

LIFE-HISTORY STRATEGIES

Any root-to-branchtip path through a tree represents a possible *life history* conceived as the experience of the members of the subpopulation flowing through the tree along that path. The path doesn't describe the life history in full detail, but it is complete with respect to the traits and environmental variables pertinent to the evolutionary problem of interest. It is natural to call the whole diagram a *life-history tree* because it is the bundle of all possible life histories with respect to the characters and environmental variables appearing in the tree. The tree structure is essentially a pictorial way of organizing and generalizing certain elements of traditional evolutionary life-history theory.

(*Caution*: The terms *life history*, *life-history strategy*, and so on are used here with meanings that may not correspond exactly with their usual meanings in the evolutionary literature. Life-history traits are conventionally conceived as special sorts of traits such as size at birth, age at maturity, and so on [Stearns 1976; 1992, 10]. Here a trait in a life-history tree need not be a character of a special sort, for any heritable phenotypic character consistent with the overall modeling assumptions can be treated diagrammatically as a life-history trait. Moreover, it is not obligatory that the events in a tree appear in a left-to-right order reflective of their chronological order of occurrence throughout the season, as the phrase 'life history' might misleadingly imply. Simultaneous or temporally overlapping events can be accommodated in the tree diagrams, as can those whose temporal order is unknown or different for different individuals. The probabilities in the trees are conditional probabilities, each probability being conditioned on whatever precedes it along the life-history path on which it lies. The ordering of nodes along paths need only accord with the logical order of these conditionalizations.)

If at each square node of a life-history tree all but one of the exit paths is pruned away, the result is a branching structure called a *life-history strategy*. Figure 2.5a shows one possible life-history strategy obtainable by pruning the earlier tree. It is the strategy of having a shell and fleeing if a predator arrives. The numbers over the nodes are computed by applying the earlier right-to-left algorithm to this strategy branch alone. The number calculated for the root node is then the rate of increase for a hypothetical population all of whose members adhere to this particular strategy. The strategy in Figure 2.5a is distinctly

33

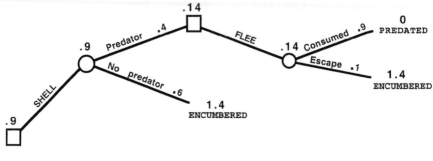

(a) A suboptimal strategy

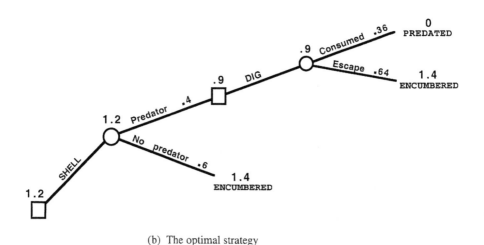

(b) The optimal strategy

Figure 2.5. Examples of life-history strategies. (a) The disadvantaged strategy SHELL + FLEE. (b) The advantageous strategy SHELL + DIG.

unpromising because the R-value over the root node is only 0.9, not enough for a population following the strategy to sustain itself.

Similar diagrams could be constructed to represent the three other life-history strategies extractable from the tree of Figure 2.4. They are the strategies associated with the remaining three trait combinations SHELL + DIG, NO SHELL + DIG, and NO SHELL + FLEE. Of these, the life-history strategy for SHELL + DIG is of special interest because, as demonstrated by Figure 2.4, it is the one selected for. It is shown in Figure 2.5b, which was obtained from Figure 2.4 by pruning at the double lines.

CONDITIONAL STRATEGIES

For a more interesting example of a life-history strategy, suppose the organism is capable of sensing quickly whether it is on soft sand or harder ground as a predator approaches. Rather than having just a simple instinct for either digging or fleeing, the organism might have a conditional response, exhibiting a different behavior depending on whether it finds itself on a soft or a hard patch when the predator arrives. An especially promising arrangement would be one in which an individual discovered on soft ground digs, whereas one on hard ground flees. A strategy involving this conditional behavior is shown in Figure 2.6.

A life-history strategy of this sort is a *conditional strategy* because it involves a character whose values are dependent upon environmental conditions. Conditional strategies of various kinds can be represented by elaborating such strategy diagrams. Nothing essentially new is involved here except the idea that because of conditional behavior or some other manifestation of intragenerational phenotypic flexibility, the selection of a character may be dependent upon the values of environmental variables preceding it in the tree.

Notice that there is not necessarily a one-to-one correspondence between trait combinations and conditional life-history strategies. That point is demonstrated by the example just given (Figure 2.6) in which the strategy in question fails to correspond to any single one of the four possible primitive trait combinations. A conditional life-history strategy is something subtler than a simple trait combination because it can conditionally involve more than one combination.

In the most general formulation of life-history strategy theory it is conditional life-history strategies that are selected for or against. The targets of natural selection are not just traits, nor even trait combinations, but entire conditional strategies. Some strategies are special cases reducing to straightforward trait combinations, but many conditional strategies are not. Any theory of selection that fails to accommodate this fact – that it is in general strategies rather than traits or trait combinations that selective pressures bear upon – is seriously incomplete.

This is merely a diagrammatic way of making a point that has been made repeatedly in the evolutionary literature. The point is usually expressed by saying that to understand the operation of natural selection in all its phenotypic manifestations one has to take proper account of heritable phenotypic *plasticity* (or 'flexibility', etc.). Plasticity refers

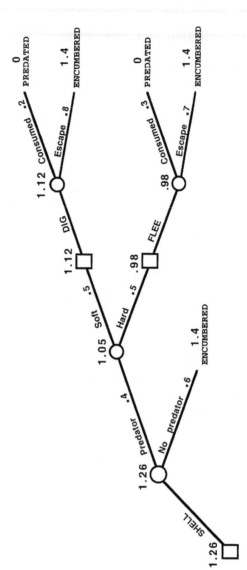

Figure 2.6. A conditional life-history strategy.

to the dependency of phenotypic properties on environmental conditions within an individual lifetime, and is manifested in such phenomena as norms of reaction, dependent development, developmental switches, conditional behaviors, or other heritable responses to environmental events (Godfrey-Smith 1996; Stearns 1989).

FITNESS

The *fitness* of a character or genotype is commonly measured by the rate of increase of the corresponding population (e.g., Hartl 1980, 201). In Model 1 this is just the magnitude of R. Some authors, including myself, have grave reservations about using populational growth rates as universal measures of fitness. Nevertheless in Model 1 it seems a harmless enough practice so long as one doesn't ask too many questions. More on this later after other models have been introduced.

Provisionally accepting the growth-rate definition of fitness as adequate for Model 1, we find that it meshes neatly with the theory of life-history trees. With the R-values interpreted as fitnesses, all of the numbers written above the consequences and nodes can be interpreted as fitnesses of the associated subpopulations. The life-history tree method as a whole emerges as a fitness maximization procedure – a diagrammatic vehicle for fitness optimization. The trait, trait combination, or conditional strategy that is selected for according to the tree procedure is always the fittest, and the fittest is always the one selected for. This happy equivalence is in fact the justification for calling R a fitness measure for the model.

FITNESS SCALES

There is an important technical point to be made about the scaling of fitness. R is acceptable as a measure of fitness in Model 1 because measuring fitness by R makes the maximally fit strategy the one that wins the evolutionary races. Using R as fitness makes the tree procedure work as a biologically justified fitness-optimization device. But might there be alternative ways of measuring fitness that also make the tree procedure work?

Actually there are infinitely many such measures. Each of the consequences appearing at the branchtips of a life-history tree has a certain value of R associated with it. This assignment of R-values to consequences leads via the tree computations to the determination of a certain character or conditional strategy as the one favored by natural selection. Now suppose all the branchtip R-values are subjected to some positive linear transformation, and the tree computations are repeated using the transformed values. It is mathematically guaranteed that the end result will be identical – the same trait or strategy as before will be found to be the advantaged one. (A positive linear transformation is a function defined by a formula of form $y = a x + b$ where a and b are real constants and a is positive.)

So we are not stuck with R as the only appropriate fitness measure for Model 1. We could equally well choose any positive linear transformation of R and the conclusions drawn from the tree procedure would remain the same. To illustrate, suppose the problem data of Figure 2.4 were arbitrarily transformed by applying the formula $y = 10R - 5$ to the branchtip R-values. Under this new choice of fitness scale the consequence assignments become

Consequence	Fitness
UNENCUMBERED	10
ENCUMBERED	9
PREDATED	−5

The tree diagram as a whole would change correspondingly to the new one shown in Figure 2.7. There the figures over the nodes no longer register true populational flows, but the tree method nonetheless correctly indicates which trait is advantaged. The life-history strategy SHELL + DIG is still the one selected for.

The situation can be summarized this way. For the purpose of using the tree method to find the direction in which the selective pressures operate in Model 1, *quantitative fitness is determined only up to a positive linear transformation.* Which particular positive linear transform of R to use in defining fitness is an arbitrary matter and the choice can be made on grounds of convenience. The case is akin to that of measuring temperature, where one can normally adopt either the Centigrade or Fahrenheit scale with equal scientific validity.

Using the measure R itself to quantify fitness is one choice of scale, the usual one. It amounts to adopting the identity transformation for scaling purposes. Beside being convenient, it has the intuitive advantage of providing a biologically meaningful zero point, because $R = 0$

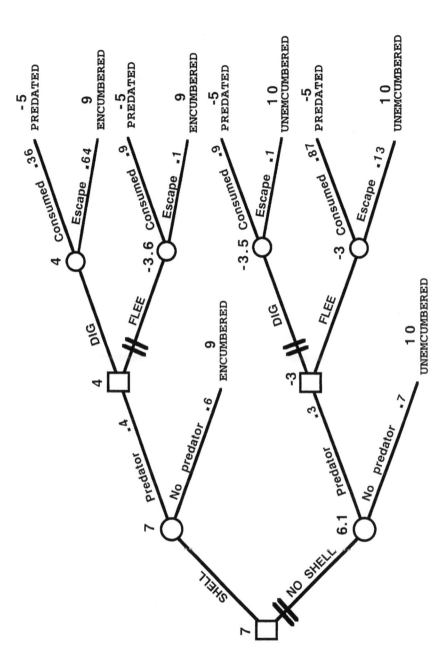

Figure 2.7. The tree of Figure 2.4 using the alternative fitness measure $10R - 5$.

signifies immediate extinction. In this respect R is comparable to the Kelvin temperature scale, which also has a physically meaningful zero point (absolute zero). But where a physically meaningful zero point is not required, as in Model 1 tree construction, other scalings can be used.

In the received statistical terminology, R (like Kelvin temperature) is said to be *ratio* measurable. Positive linear transforms of it are only *interval* measurable. The point to remember is that interval measurability is all that is needed to make the tree procedure work. The tree analyst is free to adopt any fitness scale in Model 1 so long as it differs from R by no more than a positive linear transformation.

ADAPTATIONIST REASONING

Fitness optimization reasoning is often used (though hazardously) to predict that certain characters or strategies will evolve in a population; or if they have already been observed, to explain why they appeared. The tree apparatus facilitates the construction of scientific predictions and explanations of this sort. Suppose, for instance, it has been observed that the strategy of maximal fitness in some tree has become fixed in the population. The tree diagram then becomes a possible scientific explanation of that fact. If, independently of the tree diagram, there is also historic evidence that the organism's evolutionary trajectory passed from a suboptimal strategy to the optimal one, the reasoning implicit in the tree could justify calling the optimal strategy an *adaptation*. That term would be justified in both its connotations of current adaptive value and historical genesis through a process of fitness maximization (Gould and Vrba 1982).

As usual in evolutionary theory, caution is in order. Adaptationist reasoning has been controversial. The premise of adaptationist reasoning is that, other things being equal, whatever is fittest will occur (or be likeliest to occur). Critics of the approach are quick to point out that the 'other things being equal' clause covers a multitude of sins. For a prediction or explanation based on optimization reasoning to be reliable, one would have to assume enough evolutionary time, sufficient genetic variation, no pleiotropy or linkages between traits that would prevent optimal characters from appearing, no local optima to get hung up on, no deterrent developmental constraints, and so forth through a long list of caveats. Because of these many complications, scientific prediction and explanation would be fraught with risk if based on the tree procedure alone.

But the tree procedure is not intended to be used as a sole predictive device. There is a more modest way of describing the role of the tree method, one that escapes the antiadaptationist critique or at least puts it in perspective. What has to be remembered is that the tree deals with only one factor in what is needed for scientific prediction and explanation, namely natural selection. Fitness maximization via the tree procedure is a way of discovering the direction in which the pressures of selection act in a population, nothing more. The selective tendencies so revealed may or may not have their effect, depending upon constraints external to the domain of the tree theory itself. Seen as a specialized vehicle intended solely for the analysis of selective pressures, and contributing to scientific prediction and explanation only in conjunction with whatever other evolutionary factors may be present, the tree procedure seems reasonable enough.

Of course, a pure selection formalism would be of little interest if selection weren't important. Even the arch-critics of adaptationism Gould and Lewontin (1979) concede that "Darwin regarded selection as the most important of evolutionary mechanisms (as do we)". One need not apologize then for adopting a moderate adaptationist stance of the sort advocated by Sober who writes, "*In a population subject to natural selection, fitter traits become more common and less fit traits become more rare, unless some other force prevents this from happening*" (1998, 72, italics his). Our working policy will be to give the tree diagrams their chance to explain things.

SUMMARY

The theory of life-history strategies implicit in the tree procedure offers a powerful diagrammatic and computational means of determining which characters or strategies will be favored by processes of natural selection. As presented in this chapter it rests on the combination of simplifying assumptions called *Model 1*. The correctness of the theory within that model is presumably noncontroversial, for any population biologist would, from the same problem data and modeling assumptions, arrive in some fashion at solutions identical to those produced by the tree-diagram method, though perhaps not so conveniently. The only novel elements are the diagrammatic devices themselves with their built-in computational rules and fitness scaling conventions. The whole theory of the trees is directly derivable from ordinary population biology.

Hence life-history strategy theory as embodied in the tree-diagram method reduces to standard evolutionary theory. Under Nagel's and all other explications of reducibility, any fragment or subtheory of a theory is trivially reducible to that theory. So even if the tree method is deemed a mere diagrammatic rehash of portions of standard population biology, still the point remains that it is reducible to it. The first step up the ladder of reducibility has been taken.

3

The Evolutionary Derivation of Decision Logic

The next step up the ladder leads from life-history strategy theory to decision theory. In this chapter it will be seen that when the implications of life-history strategy trees are drawn out, they are found to lead to the classical logic of decision under uncertainty.

By way of introduction, decision theory will be reviewed as it is presently conceived by most logicians, statisticians, and economists. The theory is commonly regarded as a body of principles and techniques for describing how an ideally rational agent might make choices among available alternatives in a self-consistent or 'coherent' manner. The standard theory will be presented in the next section the way it is typically presented in textbooks written by classical decision theorists – which is to say, with no reference to evolutionary theory whatsoever.

REVIEW OF CLASSICAL DECISION THEORY

Classical decision theory is concerned with the question of how an agent might rationally choose among available courses of action. If the agent's relevant knowledge of what is so is probabilistic, the applicable theory is referred to as the theory of decision under *risk* or *uncertainty*. The latter term is preferred when the probabilities are personal or subjective. The theory of decision under uncertainty assumes that the agent is able to assign values to the possible consequences that might be experienced as a result of taking any of the various available actions. These values are called *utilities*. When the theoretical emphasis is on the interpretation of the utilities, decision theory sometimes goes under the alternative label *utility theory*.

The leading idea of the classical theory of decision or utility under uncertainty is that a rational agent will choose the act that has the highest *subjective expected utility* (Raiffa 1968). This is often referred to as Bayes' Decision Rule. The subjective expected utility of an act is defined as the probability-weighted sum of the utilities of the act's possible consequences. The probability weights are the probabilities of occurrence of the corresponding consequences if the act is chosen. Under this definition a probability-weighted sum of subjective expected utilities is itself a subjective expected utility. This gives rise to a branching structure when the computations involved in a decision problem are put in a diagrammatic form.

The theory can be taken as a theoretical account of the nature of rational or 'coherent' action. Alternatively, it can be regarded as a normative guide to how actual decision-making can be made more reasonable. In the latter case the decision-theorist attempts to analyze the decision problem, discover the agent's subjective probabilities and utilities, calculate the subjective expected utility of each course of action open to the agent, and recommend as the most rational choice the one that has maximal subjective expected utility.

To illustrate, suppose a smuggler is trying to decide whether to invest in a high-speed powerboat to replace his dinghy. The advantage of the powerboat would be that if detected and challenged by a Coast Guard cutter while carrying contraband, he could probably outrace the cutter to safety. The disadvantage of the powerboat is that, being larger and noisier than the dinghy, it is more likely to be detected in the fog and challenged in the first place. Also, the dinghy could look more innocent in case it should prove expedient to try to bluff the authorities by impersonating an innocent fisherman. The powerboat would cost a lot – $400,000 in fact – and this cost would cut into the anticipated million-dollar profit of the run.

The smuggler ponders the tradeoffs but finds it difficult to arrive at a clear judgment. A believer in thorough preparation, he calls in a management consultant who is expert in decision theory. The consultant, a person of flexible character, proves willing to oblige. After interviewing the smuggler at length about the details of the situation and analyzing the data in light of decision theory, the consultant submits his recommendation in the form of the diagram shown in Figure 3.1.

The diagram is, the consultant explains, a *decision tree*. Its square nodes represent *decision* (or 'choice') nodes. The paths emerging rightward out of the decision nodes he calls *acts* ('courses of action',

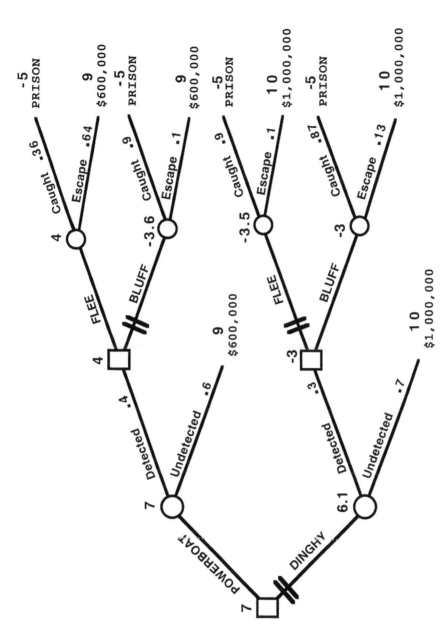

Figure 3.1. The decision tree for the smuggler's decision problem.

'choices', 'options', etc.). The round nodes he describes as *chance nodes*. The paths emerging rightward out of the round nodes represent chance *events*. After each event description a subjective probability appears. Thus, in the upper branch we see that the smuggler's subjective probability for the event of being detected by the Coast Guard is 0.4 if the power boat is bought. These probability figures represent the smuggler's own best personal estimates of the probabilities of the events in question. The consultant elicits them from him by an indirect questioning methodology that is powerful enough to reveal the smuggler's subjective probabilities even if the smuggler at first doubts he even has such probabilities.

The labels at the branchtips, in decision-theoretic terminology, describe the possible *consequences* (or 'outcomes'). The relevant consequences in this case are written as $1,000,000, referring to the possible outcome that the smuggler could make a million dollars and remain free to enjoy it; $600,000, the possible outcome that a profit of that size will remain after paying for the powerboat and he will remain free to enjoy it; and PRISON, denoting the dreary possibility of getting caught and sentenced with no monetary gain to show for the adventure.

For each of the three possible outcomes the consultant elicits from the smuggler (again by indirect methods) a corresponding subjective utility. The magnitude of each utility is written above the corresponding consequence, expressed in units called *utiles*. An assignment of utility magnitudes to consequences might be called a *valuation function*. For the smuggler's problem the valuation function is

Consequence	Utility
$1,000,000	10 utiles
$ 600,000	9 utiles
PRISON	−5 utiles

When asked why the highest utility in the table happens to be a score of exactly ten, the consultant replies that he arbitrarily set it at that round number for convenience. The scale that is chosen does not really matter, he explains, and he follows this up with a technical discourse on how utility is "determined only up to a positive linear transformation."

The final step performed by the consultant is to calculate the numbers that appear above the nodes. This is done working in leftward from the branchtips, writing above each round node the probability-weighted sum of the numbers above the nodes connected to it on its immediate right, and writing above each square node the largest of the numbers above the connected nodes on its immediate right. He describes these two computational operations as 'averaging out' and 'folding back' (Raiffa 1968). The numbers so calculated are subjective expected utilities. Act paths leading to nodes that are not of maximal utility are crossed off with a double line. They represent courses of action that are ruled out to a rational decision maker.

According to the diagram the rational course of action for the smuggler – the only way through the tree not blockaded by a double line – is to buy the powerboat and flee if detected. The consultant recommends this to the smuggler as the only strategy consistent with the smuggler's own subjective estimates of the probabilities and utilities involved. He assures the smuggler that as a consultant he has done nothing but construct an idealized representation of the smuggler's own thoughts on the subject. The decision tree is supposed to depict what the smuggler's own reasoning would or should be, in his clearer moments.

DECISION TREES AS LIFE-HISTORY BRANCHES

Compare Figure 3.1 with Figure 2.7; or more generally, compare any life-history tree with any decision tree. The resemblances are uncanny. Both are tree structures. Both are about selecting from among alternatives. Both involve chance events, probabilities, consequences, and values assigned to the consequences. The computational algorithms are identical, for both use the same averaging-out and folding-back procedures. Both kinds of tree are designed to maximize something.

In the case of Figures 3.1 and 2.7, there are not only these general resemblances but also there happens to be an exact match of particular structures and numbers. The smuggler example was contrived to have exactly the same branching structures and magnitudes as the shell example. There is no special significance to that. Any decision tree would have served to illustrate the general parallels with life-history theory. The exact match was just for fun. It is interesting though that a conventional decision problem can so easily be invented that matches the particulars of an arbitrary life-history tree, down even to the smallest structural details.

Why should human decision problems have the same structure as problems in population biology? One has two putatively independent theories, one a biological apparatus for detecting the direction in which natural selection acts on populations, the other a logical formalism for rational individual decision making. Yet problems in the one formally resemble problems in the other. Why the structural isomorphism between the two problem spaces?

Continuing the comparison, in both theories an optimal solution can take the form of a conditional strategy. Suppose for example it occurs to the smuggler that it could be a shrewd policy, if found by a Coast Guard cutter, to first observe whether the cutter is a fast or a slow model, and then bluff or flee accordingly. This conditional strategy is shown in Figure 3.2. The branch in that diagram is structurally and mathematically identical (but for a fitness scale transformation) to the conditional life-history strategy of Figure 2.6. Thus a concern with conditional strategies is another common theme linking life-history theory to decision theory.

Especially telling is the shared character of the utility and fitness scales. Although various approaches to the practical measurement of utilities have been advocated, there seems to be a clear consensus among classical decision and utility theorists on one point: Quantitative utility is determined only up to a positive linear transformation. Ratio measurability is not needed; interval measurability suffices. Thus the scale structure considered appropriate for measuring utility in decision trees turns out to be exactly what was noted in the last chapter as appropriate for measuring fitness in life-history trees. The measure-theoretic considerations are precisely parallel.

What is the significance of all these striking similarities between decision trees and life-history trees? The reductionist explanation is this. The Reducibility Thesis, as it pertains to decision theory, is reflected in the fact that *a classical decision tree is interpretable as a branch of a life-history tree*. On working out the details of life-history strategy theory, one discovers that life-history trees can have decision trees embedded as branches. Thus decision theory "grows out" of life-history strategy theory in a literal dendritic sense. It is a special case of life-history analysis.

Human decision problems are not ordinarily analyzed in this way in their full life-history tree setting, but in principle could be. According to the Reducibility Thesis, the full evolutionary analysis is implied. The foregoing remarks are only heuristic; formal confirmation of them will

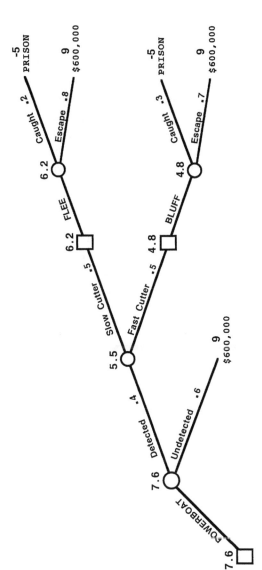

Figure 3.2. An advantageous conditional strategy for the smuggler.

49

not be forthcoming until the next chapter. Nevertheless they indicate the direction in which the reduction is headed. It is that decision theory already *is* evolutionary analysis of a special sort and needs only to be so recognized.

If decision trees are to be interpreted as branches of life-history trees there are several apparent differences to be explained. First, life-history strategy theory analyzes phenotypic characters of all types whether morphological, physiological, or behavioral. Decision theory concerns itself only with 'acts', which are exclusively behavioral. Secondly, life-history strategy trees have so far been analyzed only in terms of populational flow. Decision trees are about something seemingly different – individual choice. Thirdly, the life-history strategy trees considered so far have applied only to environmental situations that recur in exactly the same form from individual to individual throughout the population, and from generation to generation through time. Decision trees, on the other hand, can be about problem circumstances that are unique to an individual and unlikely ever to be repeated. Fourth, a life-history strategy analysis is carried out by an objective external observer of an entire population, e.g., a population biologist, while in decision theory the decision process is conceived as something carried out internally by the subjective ratiocination of an individual decision maker. Finally, regarding the quantities at the nodes and branchtips, decision-theoretic utility is not ordinarily considered to be the same thing as fitness.

If the hypothesis of reducibility is to be upheld, each of these points of apparent difference has to be scrutinized to see whether it is consistent with a reductive relationship. Putting ourselves back in the position of a logically naive evolutionist, we must see whether such an analyst might have arrived at decision theory in a natural way that accounts for these seeming divergences.

RESTRICTION TO BEHAVIORAL TRAITS

Life-history strategy theory is capable of analyzing heritable characters of any sort. Decision theory, in contrast, restricts itself to choices among acts, courses of action, strategies, etc. – in other words to behavioral traits and patterns. This is indeed a difference between the two theories, but not one that rules out a reducibility relationship. Decision theory just happens to be about those branches of life-history trees that deal exclusively with behavioral traits.

As already noted, under Nagel's and other characterizations of reducibility, any subtheory of a theory is trivially reducible to the theory. The evolutionist working in the direction of logic need be concerned only with the subtheory of life-history strategy theory that is about behavioral traits. Attention is confined to tree branches that don't involve morphological or other nonbehavioral character types. If decision theory can be shown to be reducible to that restricted subtheory, it will follow immediately that it is also reducible to life-history strategy theory as a whole.

STATISTICAL EXPECTATION

We digress for a moment to reflect on the nature of statistical expectation. Consider the following hypothetical examination question from a course on elementary statistics.

> **Problem:** A fair coin is about to be tossed. You are to receive $1 in case of heads, $0 in case of tails. What is the monetary value of this arrangement to you? Discuss.

What would constitute an acceptable answer? The question is not entirely trivial because there is no way of determining before the coin-flip takes place the monetary amount that will be received.

Two approaches are possible. First, the situation could be discussed in terms of a hypothetical *long-run average* gain. If the gamble were to be repeated many times with many independent tosses, the average gain per toss would probably be close to $0.50. The larger the number of tosses, the more exact and reliable that estimate would become.

Alternatively, the question could be answered in terms of what statisticians call *expected value.* The expected monetary value of this single future coin flip to you is exactly $0.50, calculated as the probability-weighted sum $0.5 \times \$1 + 0.5 \times \$0$. To those untrained in statistics this might seem strange because however the flip turns out you cannot receive exactly fifty cents. Rather, you will receive either one dollar or nothing and you don't yet know which. In this respect the technical statistical meaning of 'expected' differs from its colloquial meaning. But expectation is nevertheless a commonplace and highly useful notion in statistics. It provides a way of talking about the value of an isolated gamble before the outcome of the gamble is known.

An exceptionally thoughtful student trying for an A+ might answer the problem by providing both the long-run-average explanation and the expected-value interpretation of the fifty-cent figure. Such a discussion might conclude with the insight that the two ways of assessing the value of the gamble may come to the same thing. An expected value is by definition a probability-weighted average, and under at least some interpretations probabilities are themselves explicated as hypothetical long-run relative frequencies. It could be argued then that statistical propositions couched in expectational language are in principle reducible to the language of long-run averages.

POPULATION FLOW AND INDIVIDUAL CHOICE

Much of the perceived difference between life-history analysis and decision analysis may be due to the contrast between the long-run-average language used in the population-flow interpretation of the life-history trees, and the expected-value language normally used in decision theory. When the relevant biology is restated in terms of statistical expectations it immediately acquires a flavor much closer to that of decision theory.

Consider what happens when life-history strategy theory is translated into expectational terms. The first thing to be noted is the subtle change in the definition of fitness. In the population flow interpretation, fitness was defined as the factor R by which the genotypic population size increases from one season to the next. In Model 1 this is equivalent to defining it as the *average* number of offspring per population member – a long-run average. Fitness is in fact commonly defined in exactly those words (e.g., Hartl 1980, 201). They make good sense when dealing with large populations of individuals experiencing similar probabilistic processes.

But there is an alternative definition. Some authors define fitness as the *expected* number of offspring a population member will produce – a subtly different wording. This has been called the *propensity* interpretation of fitness (Mills and Beatty 1979). It has found favor because it provides a way of talking about the fitness of a single individual. Thus we find in the evolutionary literature the same duality as in statistics in general. One can with equal validity use either long-run-average language or expected-value language in defining fitness, depending on whether one is talking about a population or an individual population member.

Similar remarks apply to the numbers that appear over the nodes in a life-history tree. In the population-flow interpretation they are *average* numbers of offspring per population member, but in the expected-value reinterpretation such a number is the *expected* number of offspring for an individual. Taking a probability-weighted mean of expected values produces another expected value. The tree computation procedure therefore generates expected values everywhere throughout the tree.

In the expectational interpretation one thinks of the adventures of a single individual traversing the tree from root to branchtip. Whenever the individual encounters a square node it 'chooses' a behavior. When it encounters a round node Nature flips a coin or rolls a die for it, sending it out to one environmental experience or another. Eventually the individual comes to a branchtip consequence where it produces a certain expected number of offspring (which can be fractional!). The 'propensity' way of thinking, though mathematically no different, is conceptually distinct from imagining whole populations flowing through the tree. It is a way of thinking about an individual following a strategy. Since it is the way of thinking commonly found in expected utility theory, it provides a bridge from biology to decision theory.

In my own experience, a life-history strategy tree is something like a Necker cube illusion or the rabbit-duck visual pun. First the diagram seems to have one meaning, then after a while the mind's eye locks onto the other sense. The mental perspective shifts back and forth between the population-flow and individual-choice interpretations, between averages and expectations; yet statistically they come to much the same thing.

UNIQUE DECISION PROBLEMS

The life-history examples of the last chapter had all members of the population facing the same set of choices and circumstances. In the more general case different subsets of the population will face different variations of the decision problem. The available choices at the selection nodes may be different for different groups; the events that could ensue may be different; even if the possible events are the same their probabilities could be different; and so on. Entirely different tree

53

structures could be involved for different small subpopulations. In the extreme case, a different life-history tree diagram is needed for each individual in each generation.

Although this might seem a complication that extends beyond the reach of life-history strategy theory, in principle it can be accommodated without introducing any new theory. One does indeed have to draw a separate tree diagram for each different decision problem that an individual or group might encounter. But even if such a diagram applies only to a single individual, still it is meaningful under the expected-value interpretation. We saw earlier that it is possible to assign an expected value of $0.50 to a gamble involving equal chances of winning $1 or $0, and that statisticians consider this meaningful even if the gamble is to take place only once. In the same way, a life-history tree diagram interpreted as an individual choice analysis is meaningful even if only one individual will encounter the decision situation it describes.

Moreover, even though many different tree diagrams are needed for the many decision problems that different individuals or subpopulations may encounter, all such diagrams could (in principle) be combined into a single mammoth life-history strategy diagram. To illustrate, suppose a thousand different mutually exclusive variations on a decision situation could arise, each representable by its own tree structure. The thousand separate trees could be combined into a single inclusive tree of the sort illustrated in Figure 3.3. The root node is a chance node, and the thousand paths out of the root represent the different sets of circumstances that can arise as different variants of the decision problem that could be experienced by different individuals. Actually, a thousand is a conservative number since if continuous parameters are involved the potential variations could be effectively infinite.

When solved, the tree produces an optimal conditional strategy. It is obtained as usual by pruning off all suboptimal paths out of choice nodes. This maximally adaptive strategy could be expressed in words as one long rule: If the conditions of Decision Situation 1 arise, follow the conditional strategy that is the solution of Tree 1; If Decision Situation 2 is encountered, follow the solution to Tree 2; and so on. A genotype able to implement this many-faceted conditional strategy would be maximally fit.

There is no limit to how many variations a complex decision situation might have, and the variations need not be trivial. Considering

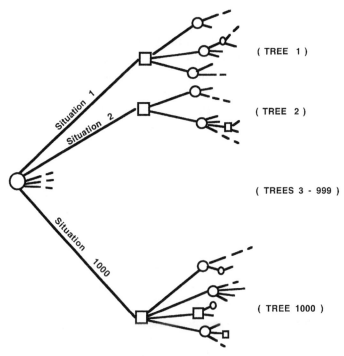

(TREE 1)

(TREE 2)

(TREES 3 - 999)

(TREE 1000)

Figure 3.3. An inclusive life-history tree comprising many problem situations.

ever larger and more finely divided problem spaces, one arrives eventually at sets of alternative decision situations that are so numerous collectively, with most so improbable individually, as to make each essentially unique to the individual experiencing it. It is mathematically obvious that when a great many mutually exclusive outcomes of a chance event are possible, with probabilities summing to one, most of those probabilities must be extremely small.

In this sense unique decision problems are already addressable by life-history strategy theory in principle. However, in practice an evolutionist would not be able to construct a diagram so immense as to analyze all the possible situations at once. Instead the analyst must be content with treating one unique decision branch at a time. That is, in fact, what is done in decision theory. A decision tree of the kind ordinarily studied in the theory of decision is one branch of a more inclusive implied life-history strategy tree.

BUSHY TREES

The implied grand life-history strategy tree may be described as *bushy* if the root is a chance node with a great many different decision situations of varied structure issuing from it as branches – branches which may themselves be bushy. The tree in Figure 3.3 is bushy, for example. On pondering the evolution of cognitive logical abilities, one comes to recognize that much depends on bushiness.

To see why, consider first a tree that is *not* bushy. Suppose only two or three branches emerge from the root, each with substantial event probabilities, and these branches are themselves nonbushy. In such a tree, how might a behavior arise that maximized fitness by following the optimal strategy? A simple instinct will do it. If there are only a few paths through a tree, each fairly probable, significant numbers of the population will follow each path generation after generation. There will be opportunity for a genotype for the optimal strategy to increase faster than its competitors and take over the population. Hence an innate behavior enacting the optimal strategy could be expected.

Such a strategy may be conditional, and even moderately complex if there is extended branching, but it will still be only a matter of automatic responses to the limited circumstances dealt with in the tree. The essential requirement is an ability to use sense data to distinguish among the few decision situations emerging from a decision node. This allows the individual to tell which decision situation it finds itself in and respond by bringing into play the appropriate instinctive behavior. Importantly, there is no need for the individual to understand or analyze the decision situation in any way. It need only detect instinctively which situation it is in by sensing some characterizing environmental clue. Then it can react, blindly, to do what it has been programmed to do.

The instinct that arises to accomplish this could be compared to a genetically bestowed stored table. The individual looks up in the left column of the table the situational clue that it has detected, and then applies the strategy found on the right. The strategy that is found, if conditional, may itself call for further situation-detections and further lookups, and these still further lookups, through any depth of nesting. Because the nesting could get somewhat complicated the usual phrase 'simple instinct' doesn't seem adequate. However, with the lookup metaphor in mind, one might aptly speak of this kind of instinct for dealing with particular situations as a *lookup instinct*. An information processing capacity sufficient to accommodate iterated lookups is all that is needed to support the evolution of optimal strategies in non-bushy trees.

Now consider bushy trees. If a tree is bushy enough, the foregoing reasoning breaks down. For as one contemplates bushier and bushier trees, there comes a point where the number of decision situations emerging from the root is so great, and the probability of most of the separate situations so low, that significant numbers of population members will no longer flow along each path. Along most branches the population flow will become too tiny a trickle for natural selection to be able to work. In the extreme case each decision situation will be essentially unique to the individual encountering it, with no ancestor having ever experienced its like. In a sufficiently bushy tree a lookup-type adaptation for the optimal decision strategy simply has no opportunity to evolve.

What will happen instead? There will be selective pressure in the direction of a more sophisticated information processing capacity that enables each individual to construct a cognitive life-history tree branch appropriate to whatever decision situation it currently finds itself in. The individual must "become its own population biologist," as it were. To be fit it must perform its own life-history strategy analysis. It need never construct a whole immense life-history tree for the entire population process, for it will not need it; but it must somehow, in effect, be able to construct and solve the branch pertaining to its own choice situation of the moment, from the point in the branch at which it finds itself. There is no other way to evolve in the direction of optimality in a bushy population process.

What is involved in such an adaptation is so impressive that it would hardly seem a serious evolutionary possibility if it hadn't already occurred in some species, notably humans. For a mere lookup instinct, the individual need glean from its environment only enough clues to distinguish the decision situation it finds itself in from a few possible alternatives; the rest is lookup. But to carry out an entire tree construction for a novel situation the individual must go much farther, must process more information and in more sophisticated ways. In some sense it has to *analyze* the decision situation as a whole. It must identify the available acts and the events that might ensue from each act, events whose occurrence might in turn make further courses of action available. It must make assumptions about the probabilities of the various events, identify possible consequences, and attach fitness estimates to them. It must incorporate all these into an appropriate tree structure, which must then be solved by some procedure equivalent to the right-to-left computational tree algorithm. In sum, it has to accomplish somehow, to the extent it can, everything that a population biologist would have to do to construct and solve the branch for that situation.

Of course the organismic information processing needed to accomplish all this might not proceed in ways exactly analogous to current methods of population biology. No one supposes that an organism will literally draw trees in its brain. It has only to execute some black-box approximation of that, with the processing giving rise to behavior that looks *as if* a tree analysis had taken place. It isn't even clear that it must depend on the same general distinctions between choices, events, probabilities, consequences, and so on. The process need only result in behavior that is so interpretable to us as analysts accustomed to these concepts.

Nor need anyone suppose that the organismic processing will be perfect or complete. There are costs and tradeoffs. Sensory limitations constrain, and evolutionary time and genetic variation are not infinite. Thus an erratic and partial approximation to a capacity for correct tree manipulation is all that might evolve. Actual organismic reasoning might be only a crude proxy for a scientifically complete life-history tree-branch analysis.

The wondrous ability to perform the approximate equivalent of a tree analysis will be referred to as *logical cognition*. Though still only an instinct in the broad sense of a genetically determined behavioral constraint, logical cognition involves vastly more than a lookup instinct. In degree of elaboration and specialization it is leagues beyond lookup. Though the memory requirements could be less than for a battery of separate lookup reflexes, the processing demands are more intricate. One could expect logical cognition to coevolve with, and influence the character of, the sensory apparatus, neural capacities, and behavioral capabilities in complex ways.

Note the role of complexity. Decision-theoretic bushiness is a kind of environmental complexity – a kind that selects for logical cognition. Cognition has long been thought to be a response to complexity. Godfrey-Smith (1996, 1) expresses his 'Environmental Complexity Thesis' thus: "The function of cognition is to enable the agent to deal with environmental complexity." Logical cognition is no exception.

To paraphrase Godfrey-Smith: The function of logical cognition is to enable the agent to deal with environmental bushiness.

OBJECTIVE AND SUBJECTIVE PROBABILITIES

Suppose an organism with logical cognition is faced with a decision situation, perhaps a novel one. To be fit, it must do what it can to become its own population biologist. It must somehow accomplish the equivalent of constructing a local branch of a life-history tree, assigning probabilities to the event paths and fitnesses to the branchtips, and solving the tree. The more closely the organism's internal branch structure resembles the tree branch an omniscient onlooking population biologist would construct for it, the fitter the organism's decision behavior will be.

To clarify things let us focus for a moment on one aspect of this task, the assignment of probabilities to the event paths. Here a fundamental distinction must be made between the actual environmental probabilities and their cognitive organismic counterparts. The actual environmental probabilities are *objective* in some sense. A well-positioned population biologist could treat them as relative frequencies and gather long-run statistics on them. Alternatively, it might be possible to deduce them from physical symmetries, or from other known statistics about the environment. And even if there should not be enough data in hand for an external analyst to accomplish any of this, still in principle there do exist objective probabilities governing the decision situation. Nature is, after all, flipping coins or rolling dice of some sort – or so Model 1 assumes.

The probabilities that the organism assigns to the event paths in its internal cognitive tree have a very different status. They are *subjective* probabilities. They represent the organism's "best guesses" about the magnitudes of the corresponding objective environmental probabilities. They are like organismic hunches about the true probabilities. If there is enough relevant environmental sense data available to the organism, and the organism has evolved a sufficiently highly developed capacity for exploiting it, such guesses might even be dignified with the term 'estimates'. Caution is needed though because 'best guess', 'hunch', and 'estimate' are only *as-if* descriptions of what is going on. As usual in evolutionary theory purposive language can be misleading if not used guardedly. Certainly the organism need not be conscious that its purpose is estimation of the objective probabilities. But whatever the terms used to describe the situation, the inferred subjective probabilities will in general differ from their objective counterparts.

There are at least two reasons for the difference. One is that perfect information about the environment is seldom available to the organism, and even when it is, it may be too costly to obtain to make the obtaining worthwhile. For this reason alone the subjective probabilities will usually resemble their objective counterparts only crudely. The other reason is that the organism may not be well enough adapted to make the best use even of the few environmental clues that do happen to be available. The cognitive apparatus might not be up to it. The possible causes of the suboptimality could include all those regularly listed in evolutionary explanations of why maximal fitness is not always attained.

It could be reasonably objected that the terms 'objective' and 'subjective' are not quite what is wanted here. Under exceptionally favorable circumstances, a sufficiently well-adapted organism with easy access to perfect environmental information could have subjective probabilities that rise to a high level of objective accuracy. In that sense they could be objective as well as subjective. To avoid this ambiguity it might be more accurate to use terms such as 'environmental' versus 'organismic', or 'external' versus 'internal', to make the distinction in question. This objection may have merit but I hope it will not cause too much confusion if we continue to use 'objective' to refer to the external/environmental probabilities and 'subjective' for internal/organismic ones.

OBJECTIVE AND SUBJECTIVE FITNESSES

Next let us consider the branchtip fitnesses of a life-history tree – that is, the numbers assigned to the consequences. Their analysis closely parallels that of the event probabilities. In particular the same distinction between *objective* and *subjective* magnitudes has to be maintained. The magnitudes that a fully informed external bioanalyst would put at each branchtip are the objective fitness figures. If a branchtip consequence has been experienced by many population members in the past, the analyst could simply use the historical average of the ensuing numbers of surviving offspring. If not, the fitness could be estimated experimentally by subjecting members of the population to the consequence and counting their offspring. And where it is not possible to take such counts in practice, still it is clear in principle that there is an objective fitness associated with each possible consequence involved in a Model 1 decision situation. The objective fitness is what is usually referred to when the term 'fitness' is used in ordinary evolutionary contexts.

The *subjective fitnesses* are, on the other hand, the magnitudes that an organism with logical cognition puts at the branchtips of its internal cognitive tree. The subjective fitnesses have a status similar to that of the subjective probabilities already discussed. If purposive language be permitted, subjective fitnesses may be said to have the character of best guesses or estimates made by the organism about the true fitnesses. They will in general differ from the objectively correct fitnesses a fully informed observer-analyst would record, both because of incomplete environmental information and because of imperfect adaptation. Consistently accurate guesswork over all consequences in all decision problems is not to be expected, though in general the better the guesses the fitter the organism.

How the subjective fitness estimates are formed is a matter for speculation, but it is hard to avoid the conclusion that knowledge of the fitnesses of at least a few of the more basic consequences must be innate. This is not an unreasonable assumption. For instance, it is not hard to believe that the extremely low fitness of the consequence PREDATED is instinctive knowledge. Other subjective fitness magnitudes, e.g., the fitnesses of novel consequences, might be learned, guessed, or estimated as expectations constructed from those already known.

UTILITIES AS SUBJECTIVE FITNESSES

If what has been suggested so far has merit, evolutionary analysis and the classical theories of choice are intimately related. Specifically, *subjective fitnesses can be identified with classical utilities*. That is to say, in Model 1 the subjective utilities that decision agents possess according to received theories of rational choice under uncertainty can be given an evolutionary interpretation as subjective fitnesses. This suggestion combines with the similar observation about probabilities – that the subjective probabilities of the evolutionary development answer well to the subjective probabilities considered in standard decision theory.

The identification of utility with subjective fitness is a leading idea in the evolutionary reduction of decision theory. Its plausibility stems from considerations ranging from the formal resemblance between decision trees and life-history trees, to the evidence for Darwinist explanations of behavior in general. Some such association has been tacitly assumed in evolutionary biology for some time, going back at least to Lewontin (1961); and many studies in evolutionary decision and game theory have in one sense or another taken fitness as a measure of utility. However, the subjective fitness interpretation of utility has not heretofore (to my knowledge) been regarded as a foundational aspect of decision theory itself.

Some traditional decision theorists might even go so far as to say that decision-theoretic utility has no connection with evolutionary fitness at all. But such an extreme position is hardly tenable. There is by now a massive literature in which evolutionists and ethologists have analyzed behavior throughout the animal kingdom. If the thrust of their findings had to be summed up in a single point, it would perhaps be something like this: Despite many exceptions and complications, animals have a pervasive tendency to behave as though trying to maximize something – namely, their evolutionary fitness. There is also an extensive literature in which psychologists, economists, statisticians, decision theorists, and utility theorists have analyzed how humans behave when making choices. Their studies suggest that, despite many exceptions and complications, humans have a tendency to act as though trying to maximize something – namely, their expected subjective utility. Now consider the curious position of those who argue that utility has nothing to do with fitness. Do they really wish to maintain that humans operate on a different principle from the rest of the animal kingdom? Or that humans as expected utility maximizers are somehow different from humans as fitness maximizers?

Jeremy Bentham, a founding father of utility theory, was frank about associating utility with degrees of personal pleasure or pain (Bentham 1823). His scale of pleasure and pain could arguably be given a biological gloss as the value range of a motivating state variable internal to the organism (Cabanac 1992). But instead of following up on this biological hint, the utility theorists who succeeded Bentham chose to study only the mathematical form and functional role of utility, not its intrinsic biological character. Perhaps they had little choice, for no mature theory of mathematical population biology was available at the time. Ever since, most formal presentations of classical decision and game theory have continued to be noncommittal about what utility actually *is* (Fisher 1918). Substantive assertions about utility's own internal meaning are scrupulously avoided except for informal asides that it has something to do with 'preferability', 'desiredness', or 'wantedness'; or that it is something to be measured with a 'hedometer' (Jeffrey 1983).

The evolutionary account makes Bentham's hazy biological notion of utility more explicit. The numbers assigned to the branchtip consequences and nodes in the biological meta-analytic tree are in the first instance objective fitnesses. But they also have a secondary interpretation on an internal organismic level as utilities, definable biologically as subjective fitnesses. It is true that fitness is in general a more comprehensive notion than utility insofar as it is applicable to organismic characters of any sort, whereas utility pertains only to behavioral acts. Also, the propensity definition of fitness must be adopted before fitness and utility become comparable. But in contexts in which these understandings are in place, and it is subjective rather than objective fitness that is of concern, the two concepts merge.

Once this point has been recognized, any perceived difference between subjective fitness and utility becomes a mere difference in point of view. The values at the tree branchtips and nodes take on different appearances from different locations on the ladder of reducibility. Seen from beneath – that is, seen evolutionarily in an internal organismic perspective – they look like subjective fitnesses. From above, the decision-theoretic view, they look like utilities. The perspective is indeed different, but to insist that utilities are therefore distinct from subjective fitnesses would be to fall into the error of the two knights who jousted over whether the Great Shield was made of silver or gold. As it turned out, one side was silver and the other gold.

Classical decision theory has traditionally characterized utilities in terms of their functional role. Evolutionary theory now supplies the specific entity, subjective fitness, that fills the role. The orthodox decision-theorist who resists this, insisting that fitness and utility are distinct and unrelated, is really saying something like "My empty box (utility) is not the same thing as your proposed content (subjective fitness)." Perhaps not, but what the reduction indicates is that the proposed content *fits* the box. It fits the peculiar shape of the box so exactly that it is hard to avoid the inference that box and content were somehow made for each other. And if they coincide they might as well be identified. If one puts the biological notion of subjective fitness into the decision-theorist's waiting vessel of utility, one arrives at the proposed unified theory of decision *qua* life-history theory.

AN OBJECTION CONSIDERED

It might be protested that identifying utility with fitness is a major conceptual blunder. Apparent counterexamples to that identification abound. Among humans, many individuals are observed to ingest addictive substances they know to be harmful, many more indulge various behaviors known to be unhealthy, some idealists take personally unjustifiable risks, some practice contraception, voluntary celibacy is not unknown, a few even commit suicide. It would seem obvious in such cases that the subjective utilities guiding the decisions could not be fitnesses. You don't maximize your expected number of surviving offspring by becoming unhealthy, celibate, or deceased. Yet such things are done.

That could sound like a fatal objection to the theory, or indeed to any Darwinist account of human behavior, but it is not. It is based on a confusion between subjective and objective fitnesses. The putative counterexamples are merely cases in which the subjective fitnesses are especially poor counterparts of the objective. It is subjective fitnesses – personal utilities – that guide decisions, and nothing in the theory claims that subjective fitnesses always correspond perfectly with their external objective counterparts. To the contrary, the two can and usually do differ. They are related, but a general correlation is enough for the theory.

The theory says that natural selection will exert pressures in the direction of removing discrepancies between subjective and objective fitnesses. It does not, however, assert that the selection has already succeeded in removing them all. If a sudden major change in environment should come along – civilization, say – some especially wide discrepancies might well be expected to appear and linger. More generally, all the standard evolutionary reasons for why organismic characteristics can fall short of optimality are also potential explanations for why utilities, interpreted as subjective fitnesses, can differ from objective fitnesses.

COGNITIVE MECHANISMS

The reductionist notion is that organisms are selected to make their behavioral decisions as though trying to be population biologists on their own behalf. They must carry out internal activities equivalent to constructing life-history tree branches and using them to choose acts of maximal expected subjective fitness or utility. The question naturally arises as to what the specific cognitive mechanisms might be that could accomplish such a feat. No attempt will be made to answer the question here, since logical theory need not be concerned with the details of physical or psychological implementation. But it is worth pointing out that the actual internal mechanisms, whatever they may be, need not resemble in any direct way their theoretical decision tree representations.

When a decision tree is solved, the manner in which logical theorists happen to describe *what* is done should not be expected to contain the procedural details of *how* it is done. The actual tree solution algorithm need not follow the procedure the human decision theorist goes through to solve a decision tree, nor need it even be like a digital computer model of that procedure. Specifically, there are unlikely to be the counterparts of programlike representations of branching graphs, numbers associated with the nodes, registers to hold the numbers, an arithmetic unit to add and multiply the numbers, a comparison unit for telling which of two magnitudes is larger, and the like.

The reason for thinking this unlikely is that there are surely simpler arrangements that would accomplish the same thing. The adaptive process could be expected to stumble upon them first. They are hinted at by the ease with which elementary physical analogs of the tree computations can be thought up. In simple cases nothing more is needed than a primitive balance beam of the sort shown in Figure 3.4. To decide between two acts in a decision problem with with no complicated branching, one arm of the beam may be taken to represent one act and the other arm the other. Weights are hung along each arm corresponding to the utilities of the acts' possible consequences, with the distances of the weights from the center of the beam proportional to their probabilities. The beam's tilt indicates the decision with the highest subjective expected utility. Hydraulic analogs have also been suggested.

Now, if it is that easy to think up simple analog computers that would do the trick, it seems plausible that there could be neural complexes

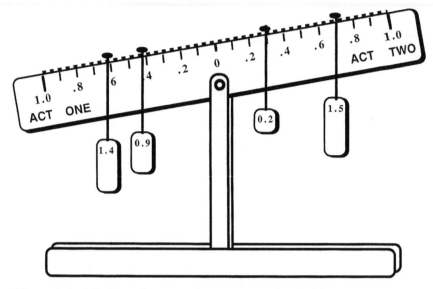

Figure 3.4. A balance-beam analog computer for decision tree computations. Distances along the beam represent probabilities, the hanging weights utilities. The magnitudes are taken from the tree diagram of Figure 2.2.

that could also solve the problem (though this would be something for a neuroscientist to confirm). It is conceivable that neural 'balance beam circuits' or their equivalent will be discovered, if they are not already known. For populations not conforming fully to the constraints of Model 1 the simplicity of the circuitry might be compromised but would perhaps still be recognizable. The details of how the circuitry accomplishes the required operations need not be the concern of logic per se.

Of course a balance beam circuit can only solve a tree once the tree has been constructed. More formidable is the cognitive task of organizing the problem tree in the first place. Mechanisms are needed for analyzing the branching dependencies and arriving at the probability and utility estimates. Many questions arise in this connection about how an organismic apparatus for information processing might best formulate consequences and events, distinguish probabilities from utilities, and so on, all assuming these entities are in fact the most efficient ways for the organism to parse its conceptual environment.

From present theory all that is clear is that for fitness sake the internal activity has to be the behavioral or 'black-box' equivalent of a decision tree diagram. Many fundamental mental constructs need to be clarified with a view to how they might be manipulated to apply the basic Bayesian decision rule. One suspects there is much grist for the mill of evolutionary epistemology in the study of trees and their efficient implementations.

SUMMARY

The similarity between classical decision trees and life-history trees is so striking that it demands an explanation. It is true that in the history of science there have occasionally been coincidental resemblances between substantively unrelated formalisms. However, in the present case there is not only the formal resemblance to consider but also the various conceptual ties. It strains credulity to believe it is all pure coincidence. Historical explanations are also unsatisfying, as evolutionary theory and decision theory developed more or less independently (until recently).

The explanation proposed here is that the logic of decision is derivable from life-history theory, even though as a historical matter it was not so derived. Decision theory results naturally from extending life-history strategy theory into the realm of individual choice behavior. The connection is subtle enough to have gone largely unrecognized, but when the two theories are compared carefully in detail the implicative relationship emerges compellingly.

To review, the chain of implications leading from evolutionary theory to decision theory runs as follows. Standard evolutionary theory includes life-history strategy theory, as seen in the last chapter. Life-history strategy theory likewise implies the subpart of itself that is confined to behavioral characters. That subtheory implies, in turn, its own translation into expected-value language, turning the life-history trees from population-flow diagrams into individual-choice diagrams. Under the latter interpretation, the theory becomes patently applicable even to choice situations that are rare or unique. Each individual can have its own private life-history tree branch for each particular decision problem it encounters.

In cases where the life-history tree or landscape is bushy, natural selection will operate to evolve a generalized capability for constructing and solving the equivalent of tree branch diagrams, a capability described here as 'logical cognition'. The hypothetical life-history analysis external to the population is realized as an actual process internalized to some extent within the individual. To the degree selection is effective, logical cognition will tend to arise in the form of an information processing capacity that allows each individual to be its own population biologist, so to speak, and solve its internal trees. Such a capacity results in the appearance of a capability for doing decision logic, with subjective fitnesses playing the role of internal utilities.

It has become commonplace in the evolutionary literature for evolutionary problems to be studied in the light of decision theory or game theory. In such studies it is often implicitly assumed, and sometimes explicitly stated, that the decision or game theory is being *applied* to the evolutionary subject matter. This seems a natural and harmless locution in the older way of thinking, but in the present context it should be taken as a warning flag that the Ptolemaic blunder may have been committed. A theory can hardly be imported and 'applied' to evolutionary theory if it is already integral to it and springs from it. Under the Reducibility Thesis the 'application' phraseology becomes inappropriate and obsolete.

4

The Evolutionary Derivation of Inductive Logic

Part I

This chapter retraces some of the steps of the last in order to reinforce the developments to this point with some formal theory. A theorem is provided to the effect that evolutionary theory gives rise to decision and utility theory as formalized by Leonard Savage (1972). The theorem shows that organisms with evolutionarily stable choice strategies conform to Savage's postulates for rational decision making. This makes decision-theoretic rationality demonstrably reducible to evolutionary considerations in a strong mathematical sense.

With the help of a second theorem it is shown that once decision-theoretic rationality has been established a theory of subjective probability is forthcoming. The Savage postulates are in fact well known as a foundation for a rigorous theory of subjective probability and utility based on choice behavior. Thus the evolutionary derivation of the Savage postulates yields not only the classical logic of decision, but also the rudiments of probabilistic or inductive logic, leading to the next rung up the reducibility ladder.

ACT REPRESENTATIONS

Since all cognitively interesting behavior can be analyzed in terms of choices among available acts or strategies, the first order of business in setting up a formal theory of decision is to settle upon mathematical representations for these entities. An abstract representation of an act is wanted that includes whatever it is about acts that is most germane to choice making.

For any problem of decision under uncertainty there is an associated set of possible *states of Nature*. A state of nature is what a biologist would think of as an environmental state or condition. The possible states of Nature with respect to a given decision problem are assumed to be mutually exclusive and exhaustive, and specified at least finely enough to take into account all the environmental distinctions needed for the analysis of the problem. One and only one of these possible states of Nature is the actual state of things. In general the decision maker doesn't know which the actual state of Nature is, which is what makes the problem one of decision under uncertainty.

The notion of a 'state of nature' (Savage's term) is not new. It is implicit wherever events with probabilities are discussed. In decision and probability theory the events relevant to a problem under analysis are all part of the so-called *event space* for the problem. In such an event space there are certain events, referred to as 'elementary events' or 'atoms', that are maximally specific in the sense that no further problem-relevant distinctions need be made within them. A state of Nature is roughly interpretable as an elementary event.

The state-of-Nature concept provides a basis for a convenient representation of acts: An act may be structured as an assignment of consequences to states of Nature. Mathematically an act then becomes a function from the set of all possible states of Nature for the decision problem into the set of all decision-relevant possible consequences. An act is a states-to-consequences mapping.

To use a favorite decision-theoretic example, suppose you are going out for a walk and must decide whether or not to pick up your umbrella as you go out the door. There are two acts available to you, to take the umbrella or not to take the umbrella. One might think offhand of representing these acts in some way that has to do with whether the umbrella goes out the door, but that would not be a problem-friendly representation. What is germane to your decision is that if you take the umbrella, you will be encumbered by it but will stay dry whether it rains or not; whereas if you don't take the umbrella, you will be unencumbered but will get wet if it rains. Thus, in the mapping representation of acts, taking the umbrella maps the event Rain to the consequence ENCUMBERED & STAY DRY, and the event No Rain to the same consequence. Not taking the umbrella is a different function mapping Rain to UNENCUMBERED & GET WET and No Rain to UNENCUMBERED & STAY DRY. A tree diagram for the acts involved in the umbrella problem has the familiar form shown in Figure 4.1. The two branches of the tree contain the mappings associated with the two acts.

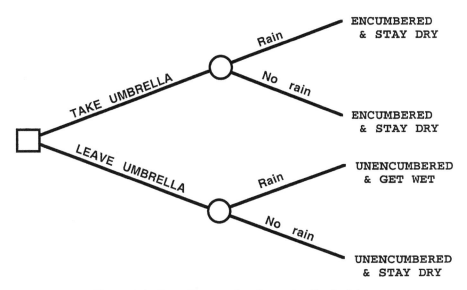

Figure 4.1. Tree diagram for the umbrella decision.

Note that the set of acts is the set of *all* possible functions from states of Nature to consequences. Typically, not all theoretically definable acts are actually available as real behavioral choices. For instance, there is (lamentably) no action you could take with regard to your umbrella that will map both Rain and No Rain into UNENCUMBERED & STAY DRY. In application, choices must be made from among actually available acts. Nevertheless it is convenient to keep all abstractly constructible acts around in the theory for the sake of completeness. They are meaningful with respect to potential behavior, for even among hypothetical acts it can still be said that one act *would* be chosen over another if the agent *were* to be confronted with a choice between them. You *would* choose an act that guaranteed UNENCUMBERED & STAY DRY for both states if only such an act were available.

Under this convention for act representation the acts assign consequences directly to states. This has the effect that in a decision tree diagram a path out of a chance node always leads directly to a consequence, never to another node. The result is a low-cropped tree structure, more of a privet hedge than a tree. The structure is illustrated in Figure 4.2 for a hypothetical decision problem involving three states of Nature s_1, s_2, s_3 and four possible consequences $C_1, \ldots, C_4$. Five

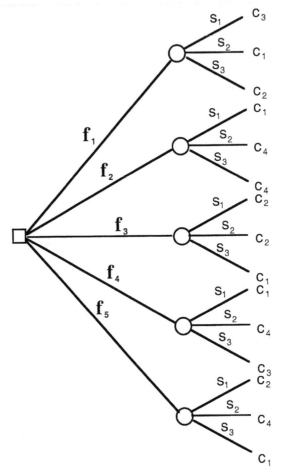

Figure 4.2. Example of a decision-tree structure using state-consequence mappings to represent acts. There is only one level of branching after the chance nodes.

available acts are shown of the $(4)^3 = 64$ acts that are theoretically constructible.

The low-cropped tree form might seem restrictive compared to the indefinitely extendible boughs that graced the trees of the previous chapter. In principle however a decision tree with long boughs can always be collapsed into an equivalent tree of the low-cut form. The reconstructed acts in the collapsed tree, described informally, may involve conditional specifications or 'if' clauses. But assuming a willingness to put up with iffy act descriptions, the reconstruction is always possible and the low-cut tree structure imposes no essential restriction on the range of applicability of the theory.

PREFERENCE RELATIONS

Preference relations between acts are a central concern of decision theory. The strict preference relation is commonly denoted $>$. If f and g are two acts, $f > g$ may be read "f is preferred to g." If neither $f > g$ nor $g > f$ obtains the two acts are said to be *indifferent*, symbolized $f \sim g$. Indifference between acts signifies that they are preferentially neutral – tied in the preference ranking; the decision agent does not care which is chosen. It is also customary to define a weak preference relation $f \geqslant g$ meaning that either $f > g$ or $f \sim g$. Decision theory is then understood to be the theory of rational ('coherent') preferences and indifferences.

Preference relations describe choice behavior. For any definite decision strategy there is a choice function C determining which act of any set of available acts is actually chosen under that strategy, or would be chosen if the set in question were to be presented to the decision maker. If the organism is confronted with a choice between two acts f and g, we write 'f C g' to signify that f would be chosen rather than g. In decision theory per se there is no mathematical definition of a choice function other than that it is a function that picks out individual acts from sets of acts. The choice function is simply a behavioral primitive that describes observable or potentially observable decision behavior.

Preference language reduces to choice language. If $f > g$ then it is understood that if forced to make a choice between act f or act g, the individual with this preference will choose to perform f. That is an inflexible requirement of the intended behavioral interpretation of preference. The possibility is specifically excluded that f might be preferred to g only in some internal psychological sense while for some reason g is what actually gets chosen. In a strictly behavioral account, preference must be defined in a way that makes preferences entirely consistent with actual choice behavior.

In endowing the preference relation with behavioral meaning, then, it is clear that if $f > g$ then f C g. A difficulty arises in connection with the converse however. When f is chosen over g, should it count as a case of strict preference (i.e., $f > g$) or a case of indifference ($f \sim g$)?

One act or the other must be chosen even when there is no partiality; indifference cannot be ruled out simply because a choice was made.

As a way of dealing with this predicament it can be required of the indifference relationship that even the slightest inducement to reverse the choice would cause the agent to change the decision. That is, $f \sim g$ may be interpreted behaviorally to mean that one act is chosen over the other but even the slightest additional inducement for the other would cause it to be chosen instead. With this understanding about what indifference means, the behavioral criterion for strict preference also falls into place. It is: $f > g$ if and only if f is chosen over g and moreover, for sufficiently slight additional inducements to choose g instead, f is still chosen.

It remains to clarify what is meant by a sufficiently slight inducement. As mentioned, the state set for a decision problem can be constructed to any desired degree of fineness. For instance a state of Nature Rain in a coarse delineation could be subdivided into two more specific states Rain & Cold and Rain & Not cold. These could in turn be subdivided further, e.g., by recognizing degrees of cold, types of rain, or by introducing entirely new factors. This sort of subdivision could in principle be repeated indefinitely to obtain a state set divided as finely as needed.

If a sufficiently fine-grained state set is assumed, there is a natural way of creating sufficiently slight inducements. "Sufficiently slight" can be taken to mean affecting the consequences on sufficiently few states. The strict preference $f > g$ can then be required to signify, behaviorally, that f is chosen over g and moreover for sufficiently small sets of states, when either f or g is modified in such a way as to map the states in the small sets into different consequences than before, the choice remains unaffected. This phrasing is only approximate; an exact definition is provided in the Appendix (Definition A.1).

THE SAVAGE THEORY

Many formalizations of classical decision and utility theory have been proposed. One survey counts twenty eight (Fishburn 1981). The most famous is the axiomatization due to von Neumann and Morgenstern (1953). Unfortunately for present purposes, their system takes probability for granted as a primitive concept – as predefined and in no need of further clarification from the theory. Since we wish to study how organismic probabilities can be constructed from more fundamental behavioral concepts, we pass their system by and take up instead what is probably the second most famous set of axioms, those of Savage (1972). Savage's system does not take probability as primitive. Instead, subjective probabilities are defined contextually from more primitive behavioral notions.

Savage's theory is a systematization of ideas related to those of Ramsey, who had earlier succeeded in constructing both personal utility and personal probability out of preferences over gambles (Ramsey 1931). The Savage theory has been criticized by some as restrictive in that it doesn't allow the states of nature to depend on the acts taken. Modifications of the theory have been proposed to overcome this perceived limitation (Jeffrey 1983; Lewis 1981; Skyrms 1984). However, it is not clear that the limitation is relevant within Model 1 and it will not be discussed further here.

Savage's formalization is consonant with the treatment of acts and preferences outlined in the previous sections. It consists of seven postulates asserting various properties a preference relation must have to be considered coherent. The posited properties are all concisely describable in terms of act structures represented as state–consequence mappings. The postulates themselves are available for study in Savage's book and other works on decision theory (Fishburn 1970, chap. 14). They are also listed for reference in the Appendix of this book, though they are not especially transparent when read in isolation from Savage's explanations.

None of the postulates makes any direct mention of either probabilities or utilities. This is a crucial methodological point. It means that when Savage derives his theory of choice from his postulates, both subjective probability and utility are defined from scratch in terms of elementary behavioral concepts. All the standard properties of probabilities and utilities are inferred from the postulated properties of preferences between acts.

From his postulates Savage is able to prove that there are such things as probabilities and utilities, at least in the sense of hypothetical entities facilitating explanations of rational decision behavior. The whole order of development of the Savage theory is one a methodologist of science could admire. The postulates are confined to what is observable or potentially observable, i.e., preferences as revealed by choice behavior. All else is inferred from the postulates, including the existence of subjective probabilities and utilities.

We shall speak of a preference relation as *Savage-rational* if and only if it satisfies the Savage postulates. The concept of Savage-rationality gives precise mathematical meaning to the notion of classically reasonable choice making.

A REDUCIBILITY THEOREM

The concept of an *evolutionarily stable strategy* (ESS) has been characterized as follows (cf. Maynard Smith and Price 1973; Maynard Smith 1984, 95):

> An ESS is a strategy that is 'uninvadable' in the following sense: If a large population consists of individuals adopting the ESS, then any mutation causing individuals to adopt some other strategy will be eliminated from the population by natural selection.

More elaborate characterizations of an ESS have been proposed, but within Model 1 the simple notion of uninvadability will be an adequate guide, sufficient to motivate the construction of more exact criteria as needed.

Viewed biologically, a choice function is a behavioral strategy. As with any other strategy it is to be expected that there will usually be selective pressure in the direction of evolutionary stability: To be fit a choice strategy must be stable. One is prompted to ask what its special structural properties might be when that condition obtains. Surprisingly perhaps, the answer is that it produces preferences that are Savage-rational.

This can be made precise. For any choice function C among acts and any weak preference relation $\geqslant$ defined in terms of C along the lines indicated,

THEOREM 4.1: If C is an ESS then $\geqslant$ is Savage-rational.

That is to say, for all Model 1 decision situations falling within the purview of Savage's theory, evolutionarily stable choice strategies automatically imply the property of rationality in Savage's sense. A proof is provided in the Appendix.

According to the theorem, rational decision making is what one should naturally expect of an evolutionarily stable product of a Model 1 process. The rationality is not something extrabiological, obedient to a supernal calculus sent down by the gods of Reason. Rather, it follows directly from the humble biological notion of evolutionary stability. Stability guarantees Savage-rationality.

This being so, rationality is something more basic than it is commonly made out to be. A tendency toward decision-theoretic rationality is predictable simply because its choice making is uninvadable, whereas a lack of rationality can be invaded. There is no need to assume that higher forms of intelligence must evolve first, after which rationality can evolve as a specialized function of some more comprehensive intellectual faculty. Rationality is not an epiphenomenon. Reasonableness in choice making is basic and stands on its own as an immediate concomitant of stability.

The 'if' clause of the theorem is an assertion in evolutionary theory; the 'then' part concerns decision theory. The postulates of decision theory are obtained from the simple evolutionary condition of stability. Because in this sense the theorem reduces decision theory to biological precepts, it seems justifiable to call it a 'reducibility theorem'. It is a reformulation of a theorem that first appeared in *Psychological Review* (Cooper 1987).

It is significant that the theorem is a theorem, not just a casual train of thought. A claim of reducibility is a claim that it would be possible in principle to construct a formal derivation of the reduced theory from the reducing. No fully formalized derivation of decision theory from elementary evolutionary theory is yet available, but Theorem 4.1 goes a long way in that direction. It lays an implicative trail from a certain point in evolutionary theory up to the Savage postulates. From there the postulates yield the classical theory of decision.

INTERPRETING THE THEOREM

The theorem says that an optimally fit preference strategy must automatically be rational in Savage's sense. This assertion is not vacuous or circular. The theorem moves from an antecedent that concerns only the organism's stability in its environment to a consequent that is solely about the internal structure of its preferences. Savage-rationality is characterized by the Savage postulates alone, and these refer only to the intrinsic structure of the preference relation and make no mention of evolutionary stability, optimality, environmental probabilities, or fitnesses.

The theorem associates rationality with stability so closely that it is well to take note of what it does *not* say. The converse of the theorem does not in general hold. Savage-rationality is a necessary but not a sufficient condition for evolutionary stability. It is possible to be rational yet unstable. Rationality alone does not ensure stability because there is more to being well-adapted than being reasonable in Savage's sense. There must, for instance, be in addition a capacity for not making poor subjective probability or subjective fitness estimates where better ones are possible; else there would be danger of invasion by strategies for making the better ones.

The theorem holds for evolutionary stability in any environment. There are no restrictions on what the objective probabilities and fitnesses imposed by the environment might be. This is highlighted when the theorem is recast in its equivalent contrapositive form:

> If a preference relation is not Savage-rational then it is evolutionarily unstable whatever the objective environmental probabilities and fitnesses may be.

Thus Savage irrationality is always selected against in every environment.

To see why irrationality always produces instability, it may be helpful to examine a particular bit of irrationality – say an intransitivity. An intransitivity is universally regarded among classical decision theorists as irrational, especially a strong intransitivity in which for certain acts f, g, and h, $f > g$, $g > h$, but $h > f$. Such cyclicality of strict preference is considered blatantly incoherent.

Though the point will hold for any state set, for simplicity suppose for the moment that the state set contains only two possible states of Nature s_1 and s_2. Picking some branchtip fitness figures arbitrarily, an intransitivity might look somewhat as shown in Figure 4.3. The three strict preferences are shown on three successive lines, each expressing a preference between two acts. The environmental probabilities of the states are left unspecified and are shown as question marks throughout.

Now let us ask, for what objective probability values p_1 and $p_2 = 1 - p_1$ will the strict preference relation $>$ be evolutionarily stable? Certainly not with p_1 close to 1.0, for the preference on the second line would then imply an unfit choice with the expected value of gamble h (near 6.0) exceeding that of g (near 1.0). The preference relation as a whole would be unstable because it could be

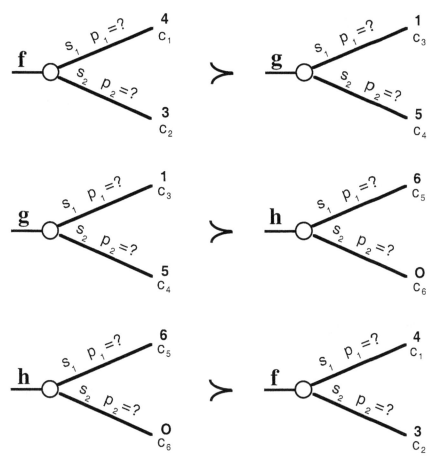

Figure 4.3. Example of an intransitivity. There is no way of assigning magnitudes to the probabilities that avoids evolutionary instability.

invaded by another preference relation identical to it except for **h** being preferred over **g**.

If on the other hand p_1 is set close to zero, both the first and the third preferences would exhibit evolutionarily instability. Continuing in this way to investigate various possible values for the probabilities, one finds that the first strict preference leads to instability if $p_1 < 2/5$, the second produces instability if $p_1 > 1/2$, and the third if $p_1 < 3/5$. Clearly there is *no* probability assignment for which this preference relation could be stable in its entirety.

There is a similar situation with respect to branch-tip fitnesses. Suppose in the diagram arbitrary probability magnitudes were filled in for p_1 and p_2 while all the branch-tip fitness values were replaced by question marks. Then it would be impossible to find fitness values for the branchtips that would not make at least one of the three strict preferences unstable. As a final exercise one could simultaneously replace all the probability values *and* all the fitness values by question marks. It would be found that there is no way of replacing all the question marks by probability and fitness values that would not give rise to instability.

This last assertion has to be qualified to the extent that there are certain choices for the probabilities and fitnesses for which there is no immediate instability among the acts **f**, **g**, and **h** themselves. A choice of values which confers an equal expectation on all three acts would be a case in point. In such situations, when decisions are made in accordance with the preferences one has neutral stability and there is invadability only by drift. Thus there is a narrow window of stability of sorts. But here the details of the definition of strict preference come into play. Satisfaction of the 'sufficiently slight inducements' clause (assuming a state set large enough to make it meaningful) implies instabilities nearby in the preference relation as a whole. Thus there is no escaping the overall preferential instability.

So an intransitivity always produces evolutionary instability whatever the objective probabilities and fitnesses imposed by the environment might be. Theorem 4.1 generalizes this insight beyond the case of intransitivities to *any* type of Savage-irrationality. Whether or not an organism has a capacity to guess probabilities and fitnesses accurately, natural selection can at least endow it with a capability for avoiding detectable preference organizations that are invadable for *all* probabilities and fitnesses an environment might impose. Irrationalities in organismic behavior are gaping chinks in the armor of stability. They are points where instability is inevitable no matter what – and it is to be expected that there will be continual selective pressures to expunge them in all environments.

THE INNATENESS OF RATIONALITY

A fit cogitator will avoid Savage irrationalities as a primate avoids snakes, and for the same reason: A tolerance of irrationalities, like a tolerance for snakes, is unstable and invadable by less tolerant strategies. Like snake avoidance, Savage rationality is likely to be largely innate, for there is no need for the intolerance to be learned anew each generation. Knowledge of the patterns to avoid can be inherited, evolving as a permanent constraint imposed on all cognitive preference structures. If this is correct and rationality is in fact essentially innate, Darwin phrased his point infelicitously when he wrote ". . . the more the habits of any particular animal are studied by a naturalist, the more he attributes to reason and the less to unlearnt instincts" (opening quote, chap. 1). *Pace* Darwin, reason probably *is* basically an unlearnt instinct.

Savage irrationalities, being structural, can be detected internally without reference to sensory input. A capacity to avoid them on that basis could be expected to evolve. This knowledge of how to avoid irrationalities could seem, to the organism, to be *a priori* knowledge, in the sense that no immediate information about the current environment is needed to obtain it. It is 'logical' knowledge that could seem to its possessor to have nothing to do with probabilities or fitnesses, and to concern only things internal or mental. The rule "Avoid intransitivities!", for instance, could seem to its possessor like a rational intuition without empirical content. Yet from a biological viewpoint such rational knowledge is full of empirical significance and has everything to do with probabilities and fitnesses. In fact such internal rules of reason presumably evolved in response to a connection with the environment of the very sort described by the theorem.

The involvement of rational coherence with objective environmental probabilities and fitnesses is disguised for the organism because for the organism they always enter into the decision logic in a universally quantified manner ("No matter what the probabilities may be . . ."). Entities that are always universally quantified need not be taken up individually, nor even recognized as individuated entities, by the cognitive system.

The theory confirms a remark made by Karl Popper. Popper writes ". . . we are born with expectations; with 'knowledge' which, although *not valid a priori*, is *psychologically or genetically a priori*, i.e., prior to all observational experience" (1963, 47, italics his). Logical knowledge can seem to the individual to be *a priori* knowledge, because it is genetically inherited. But what is *a priori* to the individual is *a posteriori* to an observer of the population process that induces the knowledge.

SUBJECTIVITY VINDICATED

Theorem 4.1 concerns objective probabilities and fitnesses through its use of the stability concept, but it does not mention subjective probabilities or subjective utilities. What are these subjective entities and how can one be sure they even exist?

There is a powerful result, due to Savage, that demonstrates that subjective probabilities and utilities *must* exist in Savage-rational organisms, in a certain sense at least. They may not be directly observable but their existence is inferable from the rational decision behavior of the organism. Coherent choice behavior is sufficient to ensure that the organism will act *as if* it had subjective probabilities and utilities, making it a metaphysical nicety whether they really exist or not. They have as much reality as many other indirect observables in the sciences.

The idea behind Savage's result is perhaps most easily grasped in terms of Bayes' Decision Rule. In thoroughly subjectivist versions of the rule it is posited that a rational agent making a choice among available acts can identify the possible consequences of each act. For each act and each of its possible consequences the agent has a certain subjective probability indicating how likely it is that the consequence in question will come about if the act is taken. There is also a subjective utility measure indicating how valuable each consequence would be to the agent if it were to come about. The *subjective expected utility* (SEU) of an act is then defined as the probability-weighted sum of the utilities of the possible consequences of the act. Bayes' Decision Rule says that a rational agent should choose the act with the greatest SEU.

Savage's result guarantees that a Savage-rational organism will behave as though in possession of subjective probabilities and utilities used in accordance with Bayes' Decision Rule. Here is an informal version of the theorem. A more exact statement is provided in the Appendix.

THEOREM 4.2: If $\geqslant$ is Savage-rational then there exists a (subjective) probability measure p over the state set and a (subjective) utility measure u over the set of all possible consequences such that for any acts $\mathbf{f}$ and $\mathbf{g}$, $\mathbf{f} \geqslant \mathbf{g}$ if and only if the SEU of $\mathbf{f}$ is greater than or equal to the SEU of $\mathbf{g}$.

Here the SEU's are understood to be computed on the basis of p and u. It can be shown further that for a given preference relation the subjective probability measure p is unique and the utility measure u is unique up to a positive linear transformation.

Theorem 4.2 can be thought of as an extension of Theorem 4.1. Theorem 4.1 asserts that evolutionary stability implies Savage-rationality. Theorem 4.2 adds that where there is Savage-rationality there must also be subjective probabilities and utilities and Bayesian choice making to make use of them. Putting the two results together, one concludes that *natural selection favors the existence of subjective probabilities and utilities and the attendant Bayesian behavior.* The subjective probabilities and utilities may or may not be good approximations of their objective environmental counterparts, but in a sufficiently stable Model 1 organism they must at least exist.

On the reducibility ladder, Theorem 4.1 formalizes the biological derivation of logic from a stage of population biology where evolutionary stability has just been introduced up to the Savage postulates. The postulates constitute the threshold of decision theory. Theorem 4.2 carries the derivation from there through classical decision theory and utility theory up to an early point in inductive logic where subjective probabilities have made an appearance. The first theorem could be considered a reducibility theorem for decision theory and the second a reducibility theorem for subjective probability and utility theory. Of course the theorems supply only the central thread of the reductions; much detail has been omitted.

MEASURING SUBJECTIVE QUANTITIES

The Savage postulates give rise, as is well known, to a theory of subjective probability whose character is governed by classical probability axioms. Savage referred to the subjective probabilities inferable from his axioms as *personal probabilities*. That terminology is also acceptable in the present context so long as no one imagines that personal probabilities are exclusive to humans. Personal or subjective probabilities and utilities are interpretable in any Model 1 population as the theoretical constructs needed to describe fit choice behavior in a bushy environment. The theorems assure their existence for evolutionarily stable organisms. Stable organisms are Savage axiom satisfiers (Theorem 4.1), and Savage axiom satisfiers act like Bayesians with coherent personal probability distributions and utilities (Theorem 4.2).

There remains the problem of measurement. Deducing the theoretical existence of subjective probabilities and utilities is one thing, measuring them in particular circumstances is another. Especially problematical is the measurement of subjective probabilities for which there is no clear physical or frequentist basis. How is one to determine Fred's subjective probability for the event that it will be raining at his house at noon tomorrow? Or Veronica's personal probability that there is life on Europa? It will not do simply to ask them, for they may not even have a clear notion of what a 'probability' is. A purely behavioral procedure is called for.

A method commonly used for eliciting subjective probabilities and utilities employs what we shall call *reference gambles*. Using reference gambles may not resolve the measurement problem completely but it does reduce it to a question involving events with clear frequentist or physical probabilities on which everybody can agree. In the smuggler's problem, the consultant might well have elicited the smuggler's subjective probabilities and utilities using this technique. The reference gambles employ chance devices with transparent physical symmetries and well-known statistical properties. Examples would include bets involving fair coins, fair dice, etc.

A chance device that is especially well-suited to the purpose would be a well-constructed roulette wheel on whose face a certain pie-shaped sector has been colored black with the rest left white. If, for example, a ninety degree sector is colored black, it becomes visually obvious that the event 'Black' should occur in about 25% of the spins. In theoretical terms that event would be said to have an objective probability of 0.25. The wheel is constructed with cleverly adjustable black and white masks so that the size of the black sector can easily be set to any fraction of the total area between zero and one.

To illustrate the use of the wheel suppose it is desired to measure Fred's subjective probability of rain at noon tomorrow. As a preliminary step two reference consequences of unequal preferability are established, say a prize of $200 and a prize of $0. Then Fred is asked to choose between two gambles. In the first, the reference gamble, the wheel will be spun and Fred will be given $200 in case of Black, $0 otherwise. In the second gamble Fred is promised $200 in case it is raining at noon tomorrow at his house, $0 otherwise. Suppose the wheel is set to probability 0.25 for Black. If Fred chooses the roulette gamble his subjective probability for rain may be deduced to be no more than 0.25. If he opts for the alternative gamble on rain, it is no less than 0.25.

After many trials of this sort using different probability settings of the wheel it will be found (assuming Fred is decision-theoretically coherent) that there is some particular setting such that for all larger settings Fred chooses the reference gamble on Black and for all smaller settings the alternative gamble on rain. At this exact setting Fred is indifferent between the two gambles. Say the indifference setting is for 0.4 probability of Black. Fred's subjective probability for rain at noon tomorrow is then declared to be 0.4. The event Black at the setting 0.4 may be described as the *commensurate reference event* for Fred for the event of rain. It is like an objective measuring stick indicating Fred's personal probability of rain.

It is noteworthy that this procedure is based solely on direct observation of actual choice behavior (or as the next best thing, the informant's word as to what he or she *would* choose). No direct mention need be made to the informant of probabilities during the elicitation, nor need the subject deal overtly with numbers in any way.

Subjective utilities can be elicited in similar fashion. Suppose a subject's utility is to be elicited for some consequence, say receiving a free ticket to tonight's concert. At a certain setting of the wheel for the probability of Black, the subject will be indifferent between receiving the ticket and a gamble paying $200 in case of Black and $0 otherwise. At all smaller settings of the wheel the ticket will be chosen, at all larger settings the gamble on Black. The event Black at this equilibrium setting is another type of commensurate reference event. The utility of the concert ticket for the subject is taken to be the same as the utility of the gamble on this reference event.

Before attempting to elicit a utility by this method, a preliminary precaution has to be taken. It must be ensured that the two reference consequences are sufficiently widely separated preferentially so that the utility of interest will fall between them. That is, one would have to check ahead of time that the subject would choose $200 over receiving the concert ticket and would choose the concert ticket over $0. If the reference consequences pass this test the reference gambles are said to *span* the problem of interest and the elicitation can proceed.

The procedure breaks down if the informant's behavior is anomalous in certain ways. Suppose, for example, that Fred chooses the roulette wheel gamble when the wheel is set to a certain probability, but the rain gamble when it is set to some higher probability. Such judgements are in conflict. When they are present there can be no commensurate reference event. This sort of situation would be an indication of decision-theoretic incoherence, and could be described as a *structural instability* in Fred's preference structure.

The elicitation procedures have their parallel in the evolutionary analysis. A structural instability is also evolutionarily unstable, it turns out. In the evolutionary context some decision problems are *recurrent* in the sense that they confront members of the population again and again, generation after generation. Others are *novel* – have seldom or never arisen before in previous generations and may be unique one-shot affairs. Evolution can solve a recurrent problem with a simple lookup instinct. An innate knowledge of the optimal choices will be selected for. A novel problem on the other hand cannot be solved by natural selection for a simple lookup instinct, because there is insufficient time and repetition for the optimal choices to evolve. Hence the analysis of novel problems is more problematical. But once at least one recurrent problem has been solved and its solution has become innate, it has the potential to be used by the organism as a reference problem for any novel problems that may arise thereafter. It supplies an internal roulette wheel and the known becomes a measuring stick for the unknown. One of the proofs in the Appendix exploits this notion of an innate objective measuring stick.

BELIEF STATES

Savage's development of probability theory from his decision-theoretic axioms will not be reconstructed here, but an informal characterization of what he ends up with can be given. Recall that the analysis began with a set of possible states of nature defined appropriately for the decision situation. If the states of nature are visualized as points in a rectangle, an event can be represented as a set of points – an area – as illustrated by event A in the Venn diagram of Figure 4.4. Other events relevant to the decision are representable as other areas. The areas for different events can overlap one another in complicated ways. The collection of all events, or subsets of the state set, is referred to as the *event space* for the decision problem.

An organism's personal probabilities with respect to a decision problem are represented by a subjective probability measure over this event space. A probability measure is an assignment of a real number

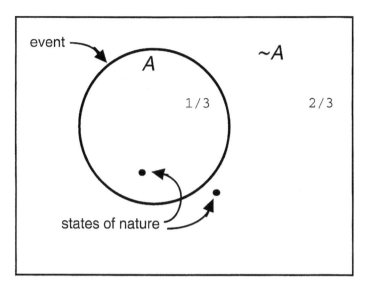

Figure 4.4. An event space with one event shown (the circular area). All points in the rectangle represent states of nature.

between 0 and 1.0 to each event in the event space, the number representing the event's probability of occurrence. For instance, in Figure 4.4 a probability of one third has been assigned to the event *A* and two thirds to the complementary event.

The theorems imply that a member of an evolutionarily stable population facing a decision problem must have an orderly subjective probability measure over the event space for the problem. Such a probability measure together with its underlying event space represents what might be called the individual's *belief state* with respect to the decision problem. The belief state reflects all the individual's current information about the relevant environmental events, digested as if in preparation for decision making. No psychological reality need be imputed to a belief state's mathematical structure. It is merely a hypothetical construct capable of explaining well-adapted decision behavior.

As an assignment of subjective probabilities to events, an organism's belief state summarizes all the individual's decision-relevant knowledge and doubts about what the world is like. One can speak of a belief state with respect to a particular local decision problem the organism faces, in which case only the probabilities of events relevant to that immediate problem will be germane. More speculatively, one could posit a comprehensive unified event space and probability assignment adequate (ideally) to cope with all potential decision problems. Such an inclusive belief state could change continually over time as new sensory information is received and the subjective probabilities are adjusted in response. In an organism exhibiting lively curiosity it could even seem as though knowledge were being collected for its own sake. In actuality however it would just be the organism's way of preparing for as wide a variety of potential future choice situations as possible, consistent with keeping the information-gathering costs affordable.

The idea of a grand all-purpose event space and probability measure is rather staggering, but it is of some comfort that Savage himself entertained a similar notion (1972, 83):

> Making an extreme idealization, which has in principle guided the whole idea of this book thus far, a person has only one decision to make in his whole life. He must, namely, decide how to live, and this he might in principle do once and for all. Though many, like myself, have found the concept of overall decision stimulating, it is certainly highly unrealistic and in many contexts unwieldy. Any claim to realism made by this book – or indeed by almost any theory of personal decision of which I know – is predicated on the idea that some of the individual decision situations into which actual people tend to subdivide the single grand decision do recapitulate in microcosm the mechanism of the idealized grand decision.

This passage echoes the idea encountered in the last chapter that there is in principle only one grand life-history tree for a population, though in practice each individual needs to confine its attention to the branch it currently finds itself on.

SUMMARY

In the last chapter it was found possible to move by a series of plausible implicative steps from evolutionary theory to decision theory, where decision theory was represented informally in terms of decision tree diagrams. In this chapter some of the same ground has been covered more formally, with proofs available in the Appendix. A well-known set of axioms for decision theory, due to Savage, has been shown to be derivable from evolutionary principles via a reducibility theorem demonstrating that evolutionarily stable preference strategies must satisfy the Savage axioms. The theorem confirms that, in Model 1 at any rate, well-adaptedness in the sense of having evolutionarily stable preferences assures decision-theoretic rationality.

The Savage postulates imply, by virtue of another theorem, the existence of subjective probabilities and utilities. They give rise to a theory of personal probability with classical formal properties. To the extent that probability theory underlies statistical reasoning, the axioms support statistical theory as well. (The title of Savage's book is, in fact, *Foundations of Statistics.*) Thus there is at least one widely received classical theory of probability and statistics that is essentially reducible to evolutionary theory.

According to some authorities inductive logic is another name for probability theory. In other usages it connotes considerably more. To whatever extent probability theory is deemed an adequate formalization of inductive reasoning, a foothold on the inductive logic rung of the reducibility ladder can now be said to have been obtained. For those who feel there should be more to inductive logic than mere probability theory and statistics, some other conceptions of induction will be taken up in Chapter 6.

5

The Evolutionary Derivation of
Deductive Logic

The ladder leads next to deductive logic, the branch of logic treating of logical implication and other logical relationships among propositions. This chapter sketches a theory of deductive logic that builds on the subjective probability theory arising from the theorems of the previous chapter. The theory is laid out in such a way as to make the deductive relationships traceable back through the probability considerations to their evolutionary roots.

The connections between probability theory and deductive logic have been variously characterized. The account of their relationship to be presented here is not necessarily the only one possible, but lends itself to a reductive approach. Much of the material is standard except for a reorganization designed to reveal the reductive architecture. Many of the underlying concepts are due originally to Rudolf Carnap (1942; 1943; 1950) Carnap and Jeffrey (1971) and Alfred Tarski (1956). However, these luminaries should not on that account be suspected of harboring any ideas about biological reducibility.

SYNTAX, SEMANTICS, AND PRAGMATICS

A logical system can be presented on any of three different levels of detail, traditionally called the *pragmatic*, *semantic*, and *syntactic* levels. Carnap, following Morris, characterized them as follows (1942, 9):

> ... If in an investigation explicit reference is made to the speaker, or, to put it in more general terms, to the user of the language, then we assign it to the field of *pragmatics*. ... If we abstract from the user of the language and analyze only the expressions and their designata, we are in the field of *semantics*. And if, finally, we abstract from the designata

also and analyze only the relation between the expressions, we are in
(logical) *syntax*.

This loose hierarchy has meant different things to different people, but
many have found it useful as a rough map of the logical territory. In
this chapter the theory of deductive logic will be developed on each of
the three levels in turn.

Traditional expositions of deductive logic usually concentrate on
the syntactic and semantic levels. The syntactic treatments deal with
axiomatics, presenting systems of logic simply as sets of axioms and
rules of inference. Semantic presentations have a more truth-theoretic
character. Propositional connectives and other devices are characterized using semantical rules of truth, truth tables, or other ways of specifying truth conditions. A valid logical implication relationship is
defined semantically by the characteristic that the consequence always
turns out to be true whenever all the premises are true.

The pragmatic level, though often neglected, will be given equal
attention here. It is important as the reductive connection to the lower
regions of the ladder. Deductive pragmatics is in fact only a short step
up from the subjective probability theory of the previous chapter. The
key pragmatic concept is that of 'belief' and the degree to which an
individual organism 'believes' a proposition. After belief has been suitably defined, a logical consequence relationship can be said to be valid
just in case any evolutionarily stable population member believing all
the premises would have to believe the conclusion too. Such a system
of pragmatics might be characterized as *belief-theoretic*, in contrast to
the truth-theoretic character of the semantical theory.

Corresponding to the three levels of abstraction, the pragmatic,
semantic, and syntactic theories of logical deduction are representable
as ascending rungs on the ladder of reducibility as shown in Figure 5.1.
The figure is a blow-up of the deductive logic rung in the original ladder
(Figure 2.1). What was earlier represented as a single rung has now
become three rungs in this finer-grained view.

All three theories of deduction yield the same results, so far as the
specific inference patterns of classical logic are concerned. In Model 1
exactly the same classical argument-forms for implicative and other
deductive logical relationships among propositions are forthcoming on
all three deductive rungs of the ladder. However, the theory that generates and explains the laws becomes increasingly abstract and parsi-
monious as the rungs are ascended.

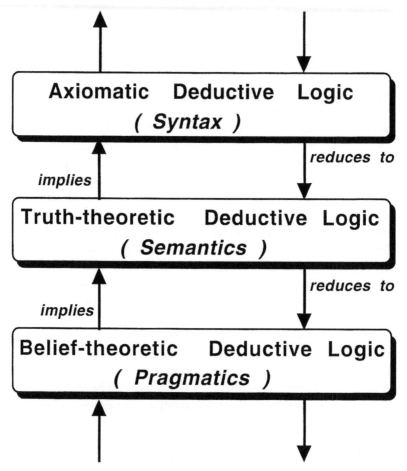

Figure 5.1. An expansion of the deductive logic rung of the original ladder of reducibility. Its pragmatic, semantic, and syntactic subrungs correspond to three successively more abstract constructions of deductive logic.

PROPOSITIONS

The subject matter of deductive logic being logical relationships among propositions, the first order of business is to arrive at an understanding of what a proposition is. In traditional theories of logic the character of propositions is frequently left obscure. In the reductive theory it is possible to do a little better.

To review, a population biologist setting out to analyze a decision problem faced by a population member does so by constructing an event space appropriate to the decision problem. From the theorems of the previous chapter the biologist knows that to the extent that the members of a Model 1 population practice evolutionarily stable choice making they will behave as though making Bayesian decisions on the basis of a probability measure in the event space. The biologist can reason freely about the events in this implied space and the inferred probabilities the individuals assign to them.

Following Carnap (1971, 35) and others we may define a *proposition* to be simply an event in an event space. In the reductive theory the space is the evolutionarily implied organismic event space. Nothing conceptually new is introduced by this identification, only a renaming. If it seems odd to equate propositions with events, it should be remembered that 'event' has a highly inclusive sense in probability theory. *The earth is round* may not sound like what would colloquially be described as an event; but it is easy to imagine mariners uncertain of the shape of the world using the locution "In the event that the earth is round . . ." and having subjective probabilities for that eventuality. Ordinarily any English expression that fits naturally into the frame "In the event that _____" names both a proposition and an event.

A proposition should not be confused with a sentence or the declaration of a sentence. The organism might or might not have a way of expressing a proposition symbolically for communicative purposes. Whether it does or not is a matter of linguistics rather than logic. The bioanalyst's symbol strings must be distinguished from any organismic linguistic utterances as well as from the propositions themselves, which are language-independent. The propositions are abstract entities in a behaviorally implied event space.

Though natural language sentences can be used to name propositions ("It rains," "There is a predator in the region," etc.), symbolic names such as **A, B, C,** . . . with subscripts added as needed are often handier in theoretical discussions. These symbolic names are *propositional constants*. In addition it is useful to introduce *propositional variables A, B, C,* and so on whose values range over the propositions in the event space. Again, such symbols are the biologist's or logician's way of referring to propositions, not the population members' way. The population members may not even have a way.

BELIEF

Let us call the subjective probability assigned to a proposition (event) in an individual's belief state the individual's *strength of belief* (degree of belief, level of belief, etc.) in the proposition. If in the belief state the event's probability is close to 1.0 then the proposition is said to be strongly believed, or to have a high credibility level. If the probability is close to zero there is strong disbelief. There are reasons for assuming, at least for the finite sorts of problems to be considered as examples here, that no ordinary factual proposition can attain a strength of belief of exactly zero or exactly one, though it could approach these values arbitrarily closely.

What then does it mean for a proposition to be 'believed'? This locution can be given an exact (though arbitrary) technical meaning by stipulating some particular numeric threshold of belief under 1.0 but close to it. A proposition is then said to be *believed* (at that threshold level) if and only if the strength of belief in the proposition exceeds the threshold.

Again there is nothing essentially new here, only the introduction of an alternative vocabulary for talking about subjective probabilities. If 'proposition' and 'belief' were introduced as primitive terms in a philosophical epistemology, they would be dubious mentalistic notions. Here however they are defined in terms of probabilistic concepts previously shown to be evolutionarily reducible. The reducible meanings of 'proposition', 'belief', etc. are not claimed to correspond perfectly with their colloquial English meanings. The English words have suggestive or mnemonic value, however.

LOGICAL CONSEQUENCE

The central concept of deductive logic is that of *logical consequence*. A logical consequence relationship (or 'logical implication', etc.) relates a set of propositions called the *premises* to another proposition called the *conclusion*. When the relationship holds, the premises are said to 'logically imply' or 'entail' the conclusion.

The pragmatic criterion of logical consequence is roughly that if all the premises are believed then the conclusion must be believed too. This is straightforward enough conceptually. However, because belief as defined in terms of subjective probability is quantitative, it needs to

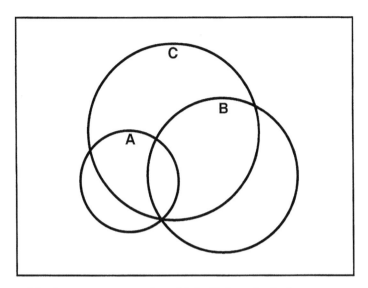

Figure 5.2. An event space in which C is a logical consequence of A and B.

be made more precise. The more detailed requirement is that no matter how high the threshold of belief is set for the conclusion, the conclusion will be believed at that level if all the premises are believed at a sufficiently high level.

DEFINITION 5.1: *B* is a (pragmatic) *logical consequence* of $A_1, \ldots, A_N$ if and only if, however high the threshold of belief for *B* may be set (short of 1.0), if each of $A_1, \ldots, A_N$ is believed sufficiently strongly then *B* is believed.

In other words a high enough subjective probability for the premises forces as high a subjective probability as you like for the conclusion. An exact mathematical definition is provided in the Appendix. The definition is patterned after definitions given by Adams (1966), Cooper (1978), and others.

To see how the definition works out in application, consider the event space of Figure 5.2. A probability measure over this space assigns a probability to each of the seven elementary sectors in the diagram in such a way that the seven probabilities add up to 1.0. The probability of any proposition is then the sum of the probabilities of the elementary sectors it includes. Note that the intersection of propositions **A** and

B is included in the proposition **C**. Hence if the probability of **A** and of **B** are both close to one, the probability of **C** must necessarily be close to one also. In belief language, no matter how high the threshold of belief is set, **C** will be believed with a strength above that threshold if **A** and **B** are both believed sufficiently strongly. Hence by Definition 5.1, **A** and **B** logically imply **C** on the pragmatic level.

In a commonly used format for displaying logical implications one writes

$$
\begin{array}{c}
\mathbf{A} \\
\mathbf{B} \\
\hline
\mathbf{C}
\end{array}
$$

where the horizontal line may be read "therefore." As a trivial application, suppose regions A, B, and C of the local rain forest are known to overlap one another in the same manner as areas **A**, **B**, and **C** in Figure 5.2. Then there could be a deduction such as

> *The predator is in region A.*
> *The predator is in region B.*
> _____
>
> *The predator is in region C.*

Because there are no possible states of nature in which the predator could be in an area of overlap that doesn't exist, the Venn diagram is valid as a map of the event space as well as of the territory.

In a valid logical consequence relationship in which the premise set happens to be empty (the case $N = 0$) the conclusion is forced by the definition to have a strength of belief of exactly one. Such a proposition is said to be *logically valid* ('universally valid', a 'tautology', a logical 'theorem', etc.) on the pragmatic level.

Logical incompatibility is another important logical relationship. Propositions are logically incompatible if they can't all be strongly believed at the same time:

DEFINITION 5.2: $A_1, \ldots, A_N$ are (pragmatically) *logically incompatible* if and only if for some sufficiently high threshold of belief (less than 1.0) $A_1, \ldots, A_N$ cannot all be believed.

That is, no probability measure over the belief state's event space can assign high probability to all the propositions simultaneously. In case $N = 1$ the lone proposition in question is said to be *logically inconsis-*

tent (or self-contradictory, tautologously false, etc.). Again a mathematically exact definition is provided in the Appendix.

PROPOSITIONAL CONNECTIVES

The bioanalyst's metalanguage can be made more versatile by introducing connective symbols that allow some propositions to be named in terms of others. Since propositions are events and events are sets of states of Nature, simple propositional connective symbols can be made directly interpretable on the pragmatic level as standard set-theoretic operations.

In the classical logic of propositions the most frequently encountered propositional connectives include conjunction, disjunction, negation, and the material conditional, commonly symbolized by **&**, $\vee$, $\neg$, and $\supset$. The traditional English renderings of these symbols are *and*, *or*, *not*, and (most controversially) *if-then*. Their definitions in terms of set-theoretic intersection, union, and complementation follow:

Connective	Usage	Definition
&	*A & B*	$A \cap B$
$\vee$	$A \vee B$	$A \cup B$
$\neg$	$\neg A$	$\sim A$
$\supset$	$A \supset B$	$\sim A \cup B$

Event names built up from propositional constants using these connectives and parentheses in the usual way are called *sentences*. They don't actually assert anything, they only name events; but the events they name are propositions that could perhaps be asserted by the organisms if the organisms had a language to assert them in

As an example of a consequence relationship involving a connective, the following reasoning holds in the event space of Figure 5.2:

$$\frac{\textbf{A \& B}}{\textbf{C}}$$

The intersection of **A** and **B** cannot be highly probable without **C** being highly probable too, fulfilling the requirements of Definition 5.1. Infinitely many other arguments are valid in the same event space though they may not be very exciting. For example the prolix

$$(A \vee A) \vee A$$
$$B \; \& \; (A \vee \neg\neg\neg B)$$
$$\overline{}$$
$$B \supset C$$

also holds.

ARGUMENT SCHEMATA

Some argument-forms are valid no matter how the events involved may be configured in the event space. One such universally valid argument schema is the venerable rule of *modus ponens*. For any two propositions *A* and *B* in any event space, it follows from Definition 5.1 that

$$A \supset B$$
$$A$$
$$\overline{}$$
$$B$$

In a passage discussing "Darwinian-selected epigenetic rules" of reasoning Michael Ruse comments "There are rules for approval of *modus ponens* and consilience no less than there is a rule setting up incest barriers. . . ." (Ruse 1986a, 161). The present theory supports this remark as it pertains to *modus ponens*, and offers the following exegesis of it. In organisms evolving under Model 1-like conditions there is selection for epigenetic rules for Bayesian behavior based on subjective probabilities in an inferred space of events or propositions. The space is structured in such a way that for evolutionarily stable individuals in which propositions $A \supset B$ and *A* are both sufficiently strongly believed, *B* must also be believed.

Many other classical inference schemata are valid under Definition 5.1. An example is the important Hypothetical Syllogism

$$B \supset C$$
$$A \supset B$$
$$\overline{}$$
$$A \supset C$$

which allows inferences to be chained.

Worthy of special mention are the so-called Paradoxes of Material Implication. The first of these is

$$\frac{\neg A}{A \supset B}$$

and the second is

$$\frac{B}{A \supset B}$$

Both these inferences are valid under Definition 5.1, just as they are in traditional developments of classical logic. Historically they have been regarded as disturbing because they seem to sanction peculiar inferences such as *The moon is not made of cheese. Therefore, if the moon is made of cheese the Republicans will win.* For the moment we note only that much of the paradoxical aspect of these inferences arises from reading the material conditional as the natural language *if-then*. It will be seen later that there are reasons for thinking that reading quite inappropriate.

Further niceties could be added to the account, but the evolutionary reduction of the standard deductive logic of propositions is now essentially complete on the pragmatic level. It will be evident to logicians that the entire classical propositional logic is derivable from Definitions 5.1 and 5.2 in the sense that all the usual classical logical consequence and incompatibility relationships among propositions follow from them. They are simply properties of the belief-state structures of evolutionarily stable Model 1 population members.

PROPOSITIONAL SEMANTICS

The ladder leads next to the semantical rung. The semantical level attempts to explain deductive logic in terms of the truth conditions governing sentences – what expressions are true under what circumstances.

In an event space, an *elementary event* is a nonempty event that has no nonempty proper subevents in the space. In a Venn diagram showing all events in an event space, the elementary events are the smallest sectors – those not further divisible. In Carnapian language an elementary event is said to represent a possible *state-of-affairs*. Conceptually a state-of-affairs is a possible combination of environmental circumstances specified at least as fully as necessary to make all distinctions needed for the decision problem.

99

Assume for expository simplicity that there are only finitely many events in the event space, that at least some of them have names (propositional constants), and that all other events can be described in terms of the named events using the propositional connectives. A conjunction of expressions each of which is either a propositional constant or the negation of a propositional constant, that contains all the propositional constants in lexicographic order, and that names an existing state-of-affairs, is called a *state description*. For example, if there are just five propositional constants A_1 through A_5, "A_1 & A_2 & $\neg A_3$ & $\neg A_4$ & A_5" would be a state description (assuming there exists a corresponding state-of-affairs). A state description serves as a sort of standard symbolic name for the state-of-affairs it denotes.

Under certain conditions a sentence is said to 'hold true' in a state description. What is meant intuitively by this locution is that the proposition named by the sentence would be true if the state-of-affairs named by the state description were the actual state of things. The holding-true relation is known as the *satisfaction relation*. When a sentence holds true in a state description the state description is said to *satisfy* (or 'fulfill', 'interpret') the sentence. The satisfaction relation can be defined recursively by the following semantical rules. For any state description,

(1) A propositional constant holds true in the state description if and only if it occurs unnegated in it;

(2) 'A & B' holds true in the state description if and only if 'A' holds true in it and 'B' holds true in it;

(3) '$A \vee B$' holds true in the state description if and only if either 'A' holds true in it or 'B' holds true in it (or both);

(4) '$\neg A$' holds true in the state description if and only if 'A' does not hold true in it;

(5) '$A \supset B$' holds true in the state description if and only if either 'A' does not hold true in it or 'B' holds true in it (or both).

(*Technical Note*: In (2)–(5) and the definition and theorem meta-schemata to follow, the variables A, B, ... are treated as sentential variables or placeholders for which particular sentences are to be substituted. The quotation marks are quasi-quotes which take effect as real quotes only after such a substitution has been made [Quine 1958, 33ff.].)

The correspondence between the foregoing semantical rules or truth conditions, and the familiar truth table definitions of the propositional connectives, should be self-evident. The truth tables are shorthand for the truth conditions. The truth-functional character that the truth tables lend to the theory is the reason for calling the semantic level of analysis 'truth-theoretic'.

The logical relationships are characterized on the semantical level as follows.

DEFINITION 5.3: For any $N \geq 0$, B is a (semantical) *logical consequence* of $A_1, \ldots, A_N$ if and only if for every state description, if 'A_1', $\ldots$, 'A_N' all hold true in it so does 'B'.

DEFINITION 5.4: For any $N \geq 1$, $A_1, \ldots, A_N$ are (semantically) *logically incompatible* if and only if there is no state description in which all of 'A_1', $\ldots$, 'A_N' hold true.

These are the counterparts on the semantical level of the earlier pragmatic definitions of the logical relationships.

The following readily proved metatheorems establish an equivalence between the semantical theory and the pragmatic:

THEOREM 5.1: B is a pragmatic logical consequence of $A_1, \ldots, A_N$ if and only if B is a semantical logical consequence of $A_1, \ldots, A_N$.

THEOREM 5.2: $A_1, \ldots, A_N$ are pragmatically logically incompatible if and only if $A_1, \ldots, A_N$ are semantically logically incompatible.

These are essentially reducibility theorems asserting that logical consequence and incompatibility, semantically defined, reduce to the comparable pragmatic relationships.

In rising from the pragmatic to the semantical level a certain simplicity is gained. There is much about classical deduction that can be studied more conveniently in abstraction from the heavy equipment of probability and decision theory. It becomes a temptation, though, to forget about pragmatic underpinnings and pretend that semantics is a science unto itself. As usual, abstraction has both its virtues and its dangers.

PROPOSITIONAL AXIOMATICS

The next rung up is the syntactic level. Here there is abstraction even from truth conditions. The logical relationships are specified by methods that involve only symbol manipulation without reference to truth or meaning. The techniques are axiomatic. An axiom system consists of a set of axioms or axiom schemata designating certain sentences as postulates, together with rules of inference for deriving some sentences from others. Axiom systems for the logic of propositions (and beyond) are provided in many works on logic and need not be reproduced here. What needs to be discussed is how they fit into the reductive account.

A *formal proof* of a sentence from a set of premise sentences is definable as a finite sequence of sentences ending with that sentence, where each sentence in the sequence is one of the axioms, or one of the premises, or follows from previous members of the sequence by one of the rules of inference. As a way of ruling out impossibilities particular to the empirical subject matter, the disjunction of all state-descriptions can be adopted as an additional axiom or implicit premise.

Given an axiomatization, the logical relationships among propositions are described syntactically by the following definitions:

DEFINITION 5.5: For any $N \geq 0$, B is a (syntactical) *logical consequence* of $A_1, \ldots, A_N$ if and only if there exists a formal proof of 'B' from 'A_1', . . . , 'A_N'.

DEFINITION 5.6: For any $N \geq 1$, $A_1, \ldots, A_N$ are (syntactically) *logically incompatible* if and only if there exists a formal proof of $\mathbf{B}_1 \, \& \, \neg \, \mathbf{B}_1$ from 'A_1', . . . , 'A_N'.

These definitions are so theory-free – so much concerned with symbols and nothing else – that a question immediately arises as to how one can know the arguments they sanction are logically valid. The answer lies in the so-called *completeness* theorems. The completeness results assert that the logical relationships defined syntactically are the same as the corresponding semantically defined relationships. Recast in a form appropriate for the present context, the standard completeness theorems for propositional logic are:

THEOREM 5.3: B is a syntactical logical consequence of $A_1, \ldots, A_N$ if and only if it is a semantical logical consequence of them.

THEOREM 5.4: $A_1, \ldots, A_N$ are syntactically logically incompatible if and only if they are semantically logically incompatible.

The proofs demonstrate that the adopted axiomatization is 'complete' in the sense that it generates all and only the semantically verifiable logical consequence and incompatibility relationships. Proofs of the completeness of various axiomatizations of classical propositional logic are available in many sources (e.g., Mates 1972). Usually they are couched in terms of 'modeling structures' or 'interpretations' that differ in form from the state descriptions employed here, but the ideas are similar.

In the reductionist theory, the completeness theorems serve as reducibility theorems. For someone who doubts that an axiomatic system of logic is scientifically meaningful, the completeness theorems reduce all axiomatic assertions to a semantical equivalent. Someone who doubted in turn that the semantics is meaningful could be shown the earlier theorems that reduce the semantics to pragmatics. And so on down the ladder to evolutionary biology at the bottom, which surely has scientific meaning.

Some methodologists have held that a theory's axioms should have a special property of "self-evidence" or "obviousness" so that there may be confidence in what is derived from them. This is not necessary in a reduced theory. In the present theory of deduction, confidence in both the axioms and what is derived from them comes up via the completeness theorems from lower logical levels and ultimately from evolutionary facts and precepts. No claim is made that the axioms and their consequent train of logical theorems represent a cognitive order of knowing within the organism. If such a claim were made for some favorite axiomatization it would be better regarded as an auxiliary psychological hypothesis than as part of logic proper.

PREDICATE LOGIC

With a three-level reductionist propositional logic in place, it remains to extend the theory beyond propositional logic to more powerful systems of classical deduction. Standard developments of first order predicate logic and quantifier theory are readily adapted to the reductive framework.

The basic picture is still that of a population biologist studying a Model 1 population. Evolutionarily stable decision strategies imply the existence of decision-relevant events which must be analyzed in relation to each other in order to understand the organismic logic. This requires that the analyst invent clever metatheoretic devices for specifying events or propositions. When the clever devices are confined to the classical propositional connectives, the theory of propositional logic is forthcoming as has been seen. To extend the theory beyond propositional logic the analyst need only introduce additional clever devices for naming or describing events.

A powerful array of descriptive devices is obtained by introducing names for objects, predicates, and relations. Relevant environmental objects or entities are denoted by *individual constants* such as $\mathbf{a}_1, \mathbf{a}_2, \mathbf{a}_3$, and so on. One-place predicate constants $\mathbf{P}_1{}^1, \mathbf{P}_2{}^1, \mathbf{P}_3{}^1$, and so on are introduced to stand for whatever decision-relevant properties these entities might have. From them *atomic sentences* can be formed by following a predicate constant by an individual constant. Thus $\mathbf{P}_2{}^1\mathbf{a}_3$ is an atomic sentence standing for the event in the belief state that $\mathbf{a}_3$ has property $\mathbf{P}_2{}^1$. Two-place relational constants $\mathbf{P}_1{}^2, \mathbf{P}_2{}^2, \mathbf{P}_3{}^2$, and so on can be used to stand for binary relations among the entities, e.g., $\mathbf{P}_3{}^2\,\mathbf{a}_4\,\mathbf{a}_1$ is an atomic sentence naming the event that $\mathbf{a}_4$ stands in the relation $\mathbf{P}_3{}^2$ to $\mathbf{a}_1$. Continuing in the same manner there can be three-place relational constants and so on. The propositional constants $\mathbf{A}, \mathbf{B}$, and so on remain available as further atomic sentences. Atomic sentences can be combined with the propositional connectives in the usual way to build up complex sentences.

With such an event-naming apparatus in place the pragmatic definitions of the logical relationships (Definitions 5.1 and 5.2) remain applicable as they stand. On the semantic level, a state description is now a conjunction of expressions each of which is either an atomic sentence or the negation of an atomic sentence, that contains all the atomic sentences in some preestablished lexicographic order, and that names an existing state-of-affairs. (This assumes the number of constants of all types is finite.) With clause (1) generalized to refer to any atomic sentence, the semantical rules remain the same as for propositional logic. The semantic definitions of the logical relations (Definitions 5.3 and 5.4) remain the same. On the syntactic level the atomic sentences play the role formerly played by propositional constants, and axiom systems for the propositional logic still apply. The definitions of syntactic logical consequence and incompatibility remain the same (Definitions 5.5 and 5.6).

Universally and existentially quantified expressions can be introduced by regarding them as abbreviations for conjunctive and disjunctive expressions involving all the individual constants. For instance, the universal quantifier ("for all x") is often indicated by putting the x in parentheses in front of an expression as in

$$(x)\!\left(\mathbf{P}_4{}^2 x\,\mathbf{a}_5\right)$$

The sentence denotes the event that all the individual objects stand in the relation $\mathbf{P}_4^2$ to $\mathbf{a}_5$. The expression may be regarded as an abbreviation for the long conjunction

$$\mathbf{P}_4^2\mathbf{a}_1\mathbf{a}_5 \quad \& \quad \mathbf{P}_4^2\mathbf{a}_2\mathbf{a}_5 \quad \& \quad \mathbf{P}_4^2\mathbf{a}_3\mathbf{a}_5 \quad \& \quad \ldots$$

where the conjunction continues until all the individual constants have been used as a first argument.

One example of the many inference schemata valid under Definition 5.1 is

$$(x)\!\left(\mathbf{P}_3^1 x \supset \mathbf{P}_5^1 x\right)$$
$$\mathbf{P}_3^1\mathbf{a}_2$$
$$\overline{}$$
$$\mathbf{P}_5^1\mathbf{a}_2$$

This is a formalization of the reasoning contained in the ancient syllogism

> *All men are mortal.*
>
> *Socrates is a man.*
>
> *Therefore, Socrates is mortal.*

It is well known that symbolic calculi of the kind just described can formalize all of the Aristotelian syllogistic and much else.

This simple system approaches the standard elementary quantification theory or *first order predicate logic* in power and expressiveness. To obtain the latter in full power, Carnap's state descriptions would have to be replaced with more versatile set-theoretic constructs called Tarskian *modeling structures*. The concepts underlying Tarskian semantics are much the same as for Carnapian, but the Tarskian modeling structures allow for infinite domains of discourse and have other technical advantages.

There are completeness theorems that generalize Theorems 5.3 and 5.4 to the first order predicate logic. The completeness proofs are available in standard presentations of classical logic. Again it should be noted that from an evolutionary perspective the completeness theorems are reducibility theorems.

In this way of looking at logic, the sentences are just event names in the bioanalyst's metalanguage. What has all this to do, then, with what the organisms are actually 'thinking'? It might be hypothesized that the organisms have some sort of private mental language of their own that in some respects resembles the analyst's language – an internal symbol system which aids them in their ratiocination. However, if such a hypothesis were made, it would be psychological rather than logical. It is not needed for the purpose of defining the logical relationships per se.

HIGHER ORDER LOGIC

It is possible to extend the first order predicate logic by the inclusion of an equality predicate and special predicate symbols for single-valued functions, resulting in what is called the first order functional calculus with equality. Going further, it is possible to introduce predicates whose arguments are themselves predicates (e.g., properties of properties, relations of relations, etc.), variables to range over them and quantifiers for those variables, still more predicates whose arguments are the earlier kinds of predicates, and so on, producing higher order logics. In the most general systems there can be properties of properties of properties, and so on, to any desired depth. Such calculi are available in both semantical and axiomatic versions. A pragmatic level is easily added so their reductive treatment becomes parallel to the logics already discussed.

For some higher order systems there is no completeness theorem. That is to say, no appropriate generalization of Theorem 5.3 or 5.4 can be proved. To the contrary, for such a system there can be an incompleteness theorem of the wicked kind made famous by Kurt Gödel. In such a system it is demonstrable that there exist semantically valid logical relationships that cannot be proved formally from the axioms and rules of inference. Nor can the axioms and rules of inference ever be extended sufficiently to repair the lack when there is essential incompleteness of that sort. Such systems are simply not fully axiomatizable and never will be. It is a limitation inherent in the axiomatic method.

This raises an interesting question. Do the incompleteness results spoil the reducibility argument for higher order logic? I think it would be more accurate to say they complicate the statement of what it is that can be reduced. What can be reduced is the metatheoretic proposition that all axiomatically derivable logical relationships are semantically valid. The converse is not reducible and not true. As a practical matter the overwhelming preponderance of ordinary semantically valid arguments are also syntactically derivable, but not all are. If the syntactic level metatheory is understood to include this qualification the reducibility ladder remains intact, and the Gödel result does not threaten the Reducibility Thesis. It merely warns that syntactic theories of logic are not always as strong and complete as one could wish.

SUMMARY

Deductive logic is ordinarily understood to be about patterns of inference in which, if the premises are known, the conclusion can be known without benefit of further factual knowledge. That is also the understanding in the evolutionary development, with provision made for the philosophically plausible precept that no factual statement can be known with absolute certainty. The provision leads to a pragmatic characterization of logical deduction in terms of degree of belief (subjective probability) and ultimately in evolutionary terms. For an organism following an evolutionarily stable strategy, a proposition that is logically implied by certain premise propositions may be seen as an event whose probability in the organism's belief state must be close to one if the probabilities of all the premise events are close to one.

Deductive argument forms can be characterized in terms of the way the propositions are described by the metaanalyst using connectives and other metadevices. If this is done using semantical rules expressing truth conditions, the argument forms are said to be valid on the semantic level. The resulting semantics is reducible to the pragmatic level by virtue of simple metatheorems that show the semantically specified valid argument forms to be identical to the pragmatic. More abstractly, one can have syntactically (i.e., axiomatically) described deductive relations. In elementary systems of deduction this level too is reducible to lower levels by virtue of theorems traditionally called completeness theorems. In reductionist logic, the completeness theorems are interpreted as reducibility theorems.

The classical systems of deduction have often been regarded as prescriptive more than descriptive. Prescriptive logic undertakes to teach reasoners how they *should* or *ought* to reason as opposed to how they might actually reason if left on their own. Is there a place for good old-fashioned prescriptive logic in the reductive picture? Perhaps there is to this extent. One could fantasize that a Model 1 population member asks a population biologist for advice about how to behave more fitly. The biologist replies, "Well, it is known that to be ideally fit, individuals in your type of population would have to behave as though managing their beliefs in conformity with this logical calculus. Here, let me teach it to you. . . ." The biologist then proceeds to inculcate the classical deductive logic as a sort of fitness therapy. Biologically, prescriptive logic is an attempt to prescribe a fitness tonic.

6

The Evolutionary Derivation of Inductive Logic

Part II

In Chapter 4 an evolutionary theory of probability was outlined as a basis for inductive logic. Probability theory itself counts as a system of 'induction' under at least some definitions of that overworked term. Other authorities have characterized induction in other ways. While it would not be practical to address all systems of induction ever proposed, in this chapter a few additional sorts of inductive reasoning will be explored. They are extensions of probability theory in significant directions that have been or could be regarded as inductive. By looking at such special kinds of induction it is possible to gain further insight into the evolutionary reducibility of inductive logic in general.

These facets of induction have been put off until now because they presuppose a certain amount of deductive theory. In terms of the ladder metaphor, it would be appropriate to imagine that a portion of the inductive logic rung has been separated off, moved up, and reaffixed as a separate rung just above the rung for the pragmatic level of deductive logic.

THE PROBABILITY CONDITIONAL

When deductive logic is built up from a platform of probabilistic pragmatics, it becomes possible to introduce sentence connectives other than the usual classical ones. An especially interesting device in this class is the conditional probability connective, or *probability conditional*. It has been studied long and fruitfully by Ernest Adams (e.g., 1965; 1975; 1998) and in his honor it is sometimes called the *Adams conditional* (Eells and Skyrms 1994; Skyrms 1984).

The probability conditional is defined by extending the definition of degree of belief. For any propositions A and B, a new expression of form $A \Rightarrow B$ is introduced whose degree of belief in a given organismic belief state is defined to be equal to the ratio of the degree of belief in A & B to the degree of belief in A. This ratio has the same magnitude as the conditional probability of ordinary statistics $Prob(B|A)$, the probability of B given A. (In the special case in which A has zero probability, ordinary conditional probability is undefined but $A \Rightarrow B$ can be arbitrarily assigned a belief value of 1.0. An interesting rationale for this assignment is provided by McGee [1994] and Adams [1998, app. 2].)

A conditional expression $A \Rightarrow B$ does not itself name any single event, so it cannot be said to denote a proposition strictly speaking. However, starting with propositions, expressions involving $\Rightarrow$ can be built up to provide ways of describing belief state properties. The simpler expressions involve just one occurrence of $\Rightarrow$ as main connective. Methods have also been proposed for extending the definitions of &, $\vee$, $\neg$, $\supset$, and $\Rightarrow$ itself to allow conditional probability connective expressions in their scope (Cooper 1978).

The probability conditional has some but not all of the formal properties traditionally expected of a conditional connective. One conventional law that it does obey is the rule of *modus ponens*:

$$A \Rightarrow B$$
$$A$$
$$\overline{}$$
$$B$$

For if the probability of B conditional on A as well as the absolute probability of A are both sufficiently high, then B must also be highly probable, justifying the inference under Definition 5.1.

The probability conditional differs in many ways from the material conditional. One welcome difference is that the probability conditional avoids the troublesome paradoxes of material implication. That is, one has

$$* \frac{\neg A}{A \Rightarrow B}$$

and

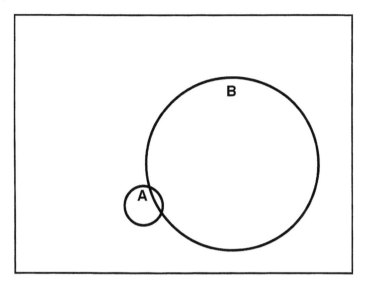

Figure 6.1. A belief state showing how the first paradox of material implication is avoided by the probability conditional.

$$* \frac{B}{A \Rightarrow B}$$

where the asterisks preceding the horizontal lines indicate that the logical consequence relationships in question do *not* hold universally.

To see how the paradoxes are blocked, consider the event space of Figure 6.1. In this kind of diagram the sizes of the areas representing the events indicate the events' probabilities. Note that the strength of belief in $\neg A$ is high (large area) yet the strength of belief in $A \Rightarrow B$ is low (low ratio of area A & B to A). Since one can make belief in the premise as strong as desired with the conclusion still unbelieved, the inference involved in the first paradox fails under Definition 5.1. A diagram is easily constructed to demonstrate that the second paradox fails as well.

The probability conditional cannot be defined by semantical rules of the ordinary sort (Adams 1975). It simply does not lend itself to a conventional truth-theoretic treatment. This may explain why its virtues are not yet widely recognized and taught in textbooks. But it has already been found capable of explaining many subtleties of common sense reasoning, and is readily accommodated within a reductive theory of pragmatics.

FORMALIZING *IF-THEN*

Natural language argumentation reflects commonsense reasoning and so is worthy of the logician's attention. The probability conditional is of special interest in that it behaves very much like the *if-then* conditional of English in its indicative usages. The avoidance of the paradoxes of the material conditional is only one of the many indications of this. One can read $A \Rightarrow B$ as "*if A then B*" almost without apology, so similar are $\Rightarrow$ and the indicative *if-then* in their logical properties.

The formalization of the indicative *if-then* by the probability conditional was suggested by Jeffrey (1964), Adams, and Brian Ellis, and has been systematized by Adams and others including myself. I have examined many patterns of *if-then* usage in informal deductive reasoning in English, and have found that the probability conditional predicts intuitively sensed logical relationships more accurately than the material conditional in almost all cases where they differ (Cooper 1978, chap. 8). In fact, for anyone willing to look closely at enough of the evidence, I think there is just no contest.

As a sample bit of evidence, Frank Ramsey observed (1931, 233) that when a person asserts *If it rains, Cambridge will win*, and another replies *If it rains they will lose*, the two conversants would ordinarily be said to have a difference of opinion. Each disputant would try to show grounds for his own belief and absence of grounds for his rival's. Any native speaker of English will agree with Ramsey that the two assertions seem to contradict each other in some sense. But when *if-then* is formalized using the material conditional, the two apparently conflicting assertions are supposed (technically) to be perfectly compatible! There is no apparent cause for dispute under that traditional formalization.

Ad hoc grounds for explaining away the sensed contradiction have been proposed, but when *if-then* is formalized by $\Rightarrow$ there is no need for ad-hocery. The conflict becomes detectable formally. Two assertions of form $A \Rightarrow B$ and $A \Rightarrow \neg B$ are always incompatible under Definition 5.2, for they cannot be believed simultaneously. Thus it is the probability conditional, not the material conditional, that is faithful to the intuitively felt *if-then* relationship in this instance.

Such examples can be multiplied. For instance, the hypothetical syllogism fails for the probability conditional:

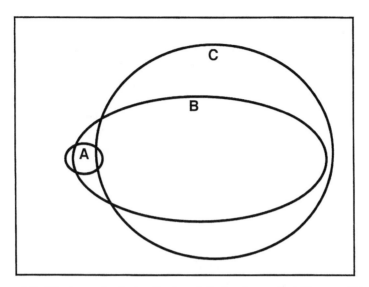

Figure 6.2. The hypothetical syllogism fails for the probability conditional because both premises could be believed and the conclusion not believed, as in this belief state.

$$B \Rightarrow C$$
$$A \Rightarrow B$$
$$*\overline{}$$
$$A \Rightarrow C$$

It fails under Definition 5.1 because it is possible for the premises to be probable with the conclusion improbable. This is demonstrated in Figure 6.2. And when English sentences are substituted that fit the conditions of the diagram, the results are indeed suspicious:

> *If I ever live in the San Francisco Bay area, I shall take up yachting.*
> *If I am ever sentenced to San Quentin prison for life, I shall be living in the San Francisco Bay area.*

* ─────────────────────────────

> *Therefore, if I am ever sentenced to San Quentin prison for life, I shall take up yachting.*

It would be possible rationally to believe both premises to a degree quite adequate for casual assertion, yet disbelieve the conclusion.

The probability conditional stands available as a useful addition to the technical metavocabulary of a biologician discussing a population's logic. Beyond that general role, if the population of interest happens to be of a species capable of expressing complicated beliefs linguistically, a linguist could hypothesize that so natural a connective as the probability conditional might be found in use among the population members themselves when they communicate. The logico-linguistic evidence concerning the indicative usage of *if-then* supports that hypothesis for the population of English-speaking humans.

PLAUSIBLE INFERENCE

To this point only static probabilities have been considered. Changes in an organism's subjective probability assignments, though recognized as frequent and normal, have not yet been examined systematically. In a well-rounded theory of induction one would like to see not only snapshots, but also movies, of belief states and how they change in response to new information. That leads into the realm of probability dynamics.

In an intriguing work entitled *Patterns of Plausible Inference*, George Polya (1954) explored a kind of reasoning that does not produce definitive conclusions but only causes propositions to become more or less plausible than they were before. He discovered many such inference patterns and he believed they play a prominent role in creative deliberation.

The following is an archetypal pattern of plausible inference:

> *A* logically implies *B*.
>
> *B* becomes believed.
>
> Therefore *A* becomes more credible.

This is intended as an argument or syllogism of sorts, but one that is understood to take place through time. There is a logical consequence *B* of *A* that is at first not believed. Then evidence of some sort is received that causes *B* to be believed. This causes *A* to take on a higher degree of belief than before. Polya's phrase 'more credible' may be taken to mean subjectively more probable.

Such inference is ubiquitous. Suppose your friend Alice is visiting abroad but you are not sure where. She had mentioned southern France and some other places in and outside of France. Obviously her presence in southern France would mean she is in France. You see from the stamp on her letter just arrived that she is in France. It immediately seems to you more probable than before that she is in southern France. The dynamic syllogism is

Alice is in southern France logically implies *Alice is in France.*

Alice is in France becomes believed.

Therefore *Alice is in southern France* becomes more probable.

The inference pattern can be visualized as taking place in the event space shown in Figure 6.3. *B* includes *A* because it is a logical consequence of *A*. At first (Figure 6.3a), neither *A* nor *B* is believed at a high level of subjective probability. Then (in Figure 6.3b) *B* comes to be believed, practically filling the diagram as its area rises to a level of belief close to 1.0. As the area interior to *B* is stretched the area of *A* within it increases proportionately, so that *A* becomes more probable.

The stretching operation depicted in the diagram makes Polya's inference pattern work as advertised. A subtle assumption is involved though. It is that the area inside proposition *B* gets stretched *evenly* when *B* is accepted. Technically what is postulated is that when an event's probability rises as *B*'s does, all probabilities conditional on the event remain the same, and similarly all probabilities conditional on the event's complement remain the same. An orderly belief state change of this sort has been described as belief revision by *probability kinematics* (Jeffrey 1983).

The plausible inference pattern may therefore be declared valid for all situations in which the probability increase in *B* proceeds by probability kinematics. It is thought that many or most probability changes do in fact proceed by probability kinematics. It is a good default assumption for use in deliberations about probability dynamics. A kinematic change is a modification prompted by making an observation that tends to confirm or otherwise change the probability of an event while giving no additional information about things not implied by the event. The rationality of belief change by probability kinematics under appropriate conditions has been defended by Skyrms (1987; 1990) and others. It is a plausible conjecture open to exploration that such arguments could be reinterpreted in such a way as to support the evolutionary fitness of kinematic change.

Another of Polya's patterns translates into belief language as follows.

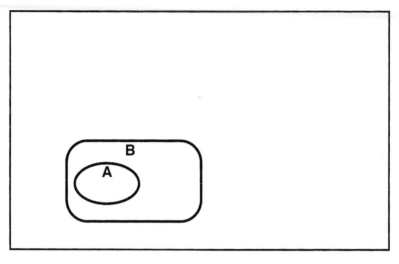

(a)   **Belief state before change**

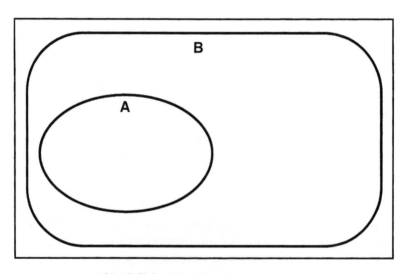

(b)   **Belief state after change**

Figure 6.3. In this pattern of plausible inference, coming to believe the conclusion of a logical implication causes a change in belief state that increases the degree of belief in the premise.

A is logically incompatible with B.

B becomes disbelieved.

Therefore A becomes more credible.

This is valid when the belief revision is by probability kinematics so that the area outside of B gets stretched evenly. When the classical connectives are introduced a profusion of further patterns can be verified, for example

$A \vee B$ becomes believed.

Therefore A becomes more credible.

It is evident that a rather rich theory of probability logic is implicit in the evolutionary probability theory that follows in the wake of the theorems of Chapter 4. The classical connectives, the probability conditional, and probability kinematics, deployed within the setting of the deductive relationships pragmatically defined, combine to provide the framework for an accommodative theory of induction and belief modification.

THE INDUCTIVE LEAP

The most familiar form of inductive reasoning, perhaps, is the kind in which there is a leap from particulars to a generality. The generalization becomes believed even though the confirming cases are far from exhaustive. In David Hume's famous example, after examining a few swans each of which turns out to be white, it is accepted (or becomes far more probable) that all swans are white.

Inductive reasoning of this sort is commonly manifest in human and animal behavior. An animal foraging in the forest comes upon a tree of an unfamiliar kind with a windfall pile of nuts under it. The animal tries one nut, finds it inedible, and drops it. A second nut is tasted, found inedible, and similarly rejected. The same for several more nuts. The animal then gives up, ignores the remaining nuts, and wanders off to forage elsewhere.

What has happened can be studied as a pattern of plausible inference. Suppose the nuts in the pile are denoted by the individual constants **a**, **b**, **c**, **d**, and so on, and that **P** is a one-place predicate signifying inedibility (subscripts and superscripts having been dispensed with for convenience). The inference is that after some of the nuts have been found inedible, it becomes much more plausible for any remaining nut in the pile that it too is inedible. The inference is made even though there is no more direct physical evidence about the remaining nuts than there was before.

Simplifying to the case where just two nuts have so far been tried and a third is in prospect, one has the pattern

> **Pa** becomes believed.
>
> **Pb** becomes believed.
>
> Therefore, **Pc** becomes more credible.

This little example sets out the inductive leap in an elementary form that invites analysis.

The inference pattern is what would be expected if the animal's prior belief state were as shown in Figure 6.4a. There, before any empirical evidence about individual nuts has been obtained, each of the three propositions (circles) **Pa**, **Pb**, and **Pc** has the same subjective probability (area). However, of the eight elementary sectors, the innermost and outermost are larger than any of the remainder. That is, the possibilities 'all-inedible' and 'none-inedible' are taken to be inherently likelier states-of-affairs than the other possible combinations of nut properties.

Figure 6.4b shows the posterior probability distribution after the two premises **Pa** and **Pb** have become believed. Notice that **Pc** has indeed become more probable than before. There has been a net stretching of that proposition – an expansion that answers to the inductive leap. Thus for the particular prior distribution that was assumed, the inference pattern is confirmed and requires no new inductive machinery.

A prior probability distribution of the sort in Figure 6.4a, in which extra probability weight is accorded to states-of-affairs exhibiting homogeneity of properties, is known as an *inductive* prior. In the obvious extension of the notion, states-of-affairs exhibiting near-homogeneity of properties (e.g., all swans white except for one or two) can be given extra weight also, though less than for complete homogeneity. In a generalized inductive prior a range of probability values can be assigned to the various states-of-affairs depending on how close each property distribution comes to complete uniformity.

We see then that this kind of inductive reasoning can be accounted for within the logical theory already at hand if an inductive prior distribution can be assumed. But how can such an assumption be justified? The short answer is that organisms tend to use inductive priors because

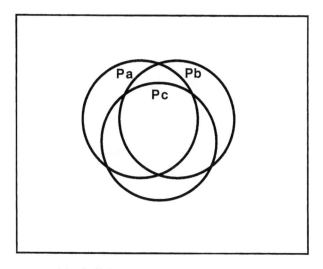

(a)   **Belief state before changes**

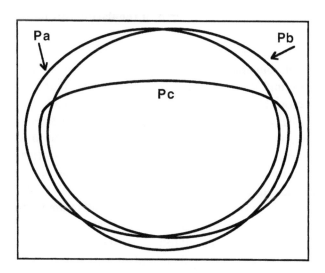

(b)   **Belief state after changes**

Figure 6.4. A plausible inference pattern in which it is learned that two objects have a certain property, causing a state change in which it becomes more probable that a third object also has the property.

the use of such priors tends to be selected for. Inductive priors are adaptive and maximize fitness. That answer suffices for an evolutionary explanation of induction. It does not necessarily solve Hume's philosophical problem of how to justify induction in any ultimate sense, but it does reduce it to one of justifying the underlying evolutionary science.

Some might scoff at the idea of the gentle creatures of the forest applying inductive priors while nibbling nuts. They may propose superficially simpler explanations for the same phenomena – associative learning procedures, for example. Such explanations could well have merit as proximate psychological mechanisms. But the mechanisms will only succeed evolutionarily to the extent that they produce the same behavioral choices as the deeper theory of inductive priors would. The ultimate analysis in terms of priors is needed to explain why the proximate procedures have survival value.

THE UNIFORMITY OF NATURE

It has long been recognized that successful Humean inference requires that the world possess some degree of regularity. This has been called the *Principle of the Uniformity of Nature*. All swans being white would be an example of such a uniformity. As it turns out, there are swans that are black; but that does not negate the principle. There is still a good deal of uniformity simply in the circumstance that swans come in only two colors, black and white. Moreover, there is additional uniformity in the heavy preponderance of just one color, white.

Suppose an organism is evolving in an environment in which there is a significant degree of uniformity. That is, the objects and natural properties in the environment that are relevant to the organism's decisions are distributed in haphazard ways, some more evenly and some less, but on average they exhibit a certain overall tendency toward uniformity. In such an environment, where there really is a tendency toward regularity, it will be adaptive for the organisms to favor the kinds of priors which reflect that tendency, namely inductive priors. Those that do so will tend to make fitter decisions on average, and will win their evolutionary races against those that don't.

It might even be possible in principle to model such an environment mathematically and calculate what the fittest priors would be for the organisms evolving under the model. The Model 1 environments of the examples to this point have involved only a few objects and properties, but in theory nothing would prevent modeling *all* environmental objects and natural properties of potential relevance to decision making. Their statistical frequencies and interdependencies could be patterned after the actual statistics of the real environment being modeled. It could then be computed what prior probability distributions would be most adaptive for organisms forced to deal with new objects and properties. With a tendency toward uniformity in the environment, the fittest priors would be inductive priors, the detailed shape of their distributions being determined by the statistics of the environment.

Such a modeling task, though provocative as a thought experiment, might not be practical as an actual scientific project because of a number of complications. One of the deeper ones is that it is not easy to say what a 'natural' property of the environment is. Philosophers have been able to specify entertainingly unnatural properties for which inductive reasoning would surely fail (the most notorious being the property 'grue' [Goodman 1955, 74ff.]). Nevertheless, even left as a gedanken experiment the notion is sufficient to suggest that a fit basis of induction for personal ratiocination must in the end be a matter of the statistics of one's environment. The proper degree and kind of inductiveness of priors is in principle reducible to scientific theory and observation. Biologically, quantifying induction reduces to a problem in fitness optimization.

It is a consequence of this view of induction that the laws of inductive logic are in general different for different populations inhabiting different environments. For some organisms priors with a high degree of inductivity would be fittest, while in other populations and environments with less homogeneity a lesser degree of inductiveness might serve better. In certain peculiar environments, oddly shaped inductive distributions could be optimal; and it is even conceivable that in some environments involving abnormally many uninductive or counterinductive property distributions, the inductive leap would not work at all. Evidently then the laws of Humean inductive logic are not fixed and universal. They can be optimal only with respect to the particular populational/environmental complexes to which they pertain.

The idea of developing inductive logic in terms of prior probability measures is not new. A notable historic attempt was made by Carnap (1950) at constructing a grand prior suitable for all inductive reasoning. Carnap's approach was to lay down *a priori* constraints that a reasonable prior ought to fulfill, and then to look for a probability measure

that satisfied them. He eventually concluded that his reasonable constraints were insufficient to determine inductive logic fully. Instead, the
reasoner is forced to select on the basis of judgment and experience
from among a "continuum of inductive methods" (Carnap 1952), or
from some even larger array of possible inductive logics (Carnap 1971,
1980). To this extent the Carnapian rules of inductive reasoning are
incomplete.

The evolutionary perspective explains and resolves the indeter-
minacy, at least in principle. Relative to any given population,
environment, and natural property class there is a unique prior
that is evolutionarily fittest. To be well adapted an organism able
to perceive those natural properties must use that prior. Though
Carnap did not frame the problem of induction in terms of a
given biological environment, he did explore methods of evaluating
the success of any particular inductive method in a given 'possible
world'. Stretching the point a little, his explorations could perhaps be
reconstrued as an unintended first step toward determining the fittest
prior for organisms inhabiting a particular biological environment.
As a life-long advocate of the unity of science, he might even have
approved.

THE ILLUSION OF ENHANCED UNIFORMITY

Analyzing induction biologically puts a curious new spin on the hoary
question of how uniform the universe really is. The objects, properties,
and relations that an organism can cognize are not a random selection
of all such entities existing in its environment. They are ones of a type
that the organism has evolved to be able to use in its decision making.
As a consequence it is to be expected that, other things being equal,
an organism's sensory and cognitive apparatus will evolve in such a
way as to favor the recognition of those types of properties that are
the most inductively distributed over the environment. Such a sensory favoring would be advantaged because inductively distributed
properties are more predictable – more amenable to inductive ratiocination by the organism. For such properties, inductive leaps succeed
more often.

Thus there is a sampling bias. The kinds of objects and properties
that a well-adapted organism has come to sense and cogitate about will
tend to some extent to be those with greater than average uniformity.

Classes of properties tending toward extreme uniformity will be especially favored. Uniformity is of course not the only aspect of properties that can affect their significance for organismic decision making, but it is one of them.

The upshot is that Nature as a whole may not be as uniform as it appears to be to well-adapted creatures such as ourselves. We might have evolved to see more tendencies toward regularity than actually exist on average. If so we may be looking at the environment as though through order-enhancing goggles. We must look to evolutionary epistemology to provide us with corrective lenses so that we may perceive more clearly the true extent of the order in the world.

SUMMARY

Because the term 'induction' has been used in so many ways, and because novel systems of induction are still being proposed, inductive logic is something of a moving target so far as evolutionary reduction is concerned. Nevertheless it can perhaps be said that the prospect of such a reduction does not seem unreasonable for the better known probability-based systems, whose basic elements are traceable back to considerations of fitness optimization.

In Chapter 4 it was shown that a standard theory of decision under uncertainty can be drawn out of evolutionary theory with probability theory as a component. Such a probability calculus is itself a kind of inductive system, or is so recognized by some authorities. In this chapter some other kinds of induction have been touched upon. One is the 'probability logic' that relates probabilities to deductive concepts through use of the probability conditional. Another is the theory of 'plausible inference' that tracks the ways in which beliefs can become more credible or less credible in accordance with certain patterns of probability kinematics. Still another is the famous Humean kind of inductive inference that moves from particulars to generalities, thought to underlie much scientific and personal reasoning. What is involved in biologically reducing these kinds of induction has now been outlined, albeit sketchily. The role of simplicity, of analogy, of event-sequence ordering, and other important topics in induction remain open to evolutionary exploration.

Lest the reductive thread be lost, what has been learned is not just that cognitively endowed organisms in a complex environment must be able to induce well to be fit, as birds must fly well to be fit. That much was already clear. Rather, the point is that the formal principles of logical induction are themselves biological principles. One could be led astray in looking for the foundations of induction elsewhere.

7

The Evolutionary Derivation of Mathematics

The final step up the ladder of reducibility leads from deductive logic to general mathematics. It is already a well-trodden path, for it has much in common with a famous position in the philosophy of mathematics known as *logicism*. The so-called *logicist thesis* asserts, roughly, that mathematics is reducible to pure logic. The doctrine is closely associated with Bertrand Russell and Alfred North Whitehead's *Principia Mathematica* (1910), widely acknowledged to be an intellectual monument of the twentieth century.

It would be impractical to present here the particulars of the logicist reduction of mathematics, and unnecessary too as so much has been written about it elsewhere. The discussion will be confined therefore to some remarks about how logicism fits in with evolutionary reductionism. The commentary will of necessity be tentative and incomplete, as this is an especially delicate stage of the reduction. It will not be suggested that this final step up the ladder has been established with finality, but only that it is worthy of consideration; and that in case the reader has already considered and rejected the logicist thesis, that it be reconsidered in the light of the reductionist program. Some of the more troubling aspects of logicism simply do not arise in the reductionist context.

LOGICISM

In the latter part of the nineteenth century the ground was prepared for a logical reduction of mathematics by such figures as Boole, Peano, and Dedekind. It was recognized that natural numbers can be described axiomatically; that signed, rational, real, and complex numbers are reducible to natural numbers; and that there is an algebra for combining propositions. A comprehensive truth-functional logic was developed by Gottlob Frege. Frege was the first to articulate the logicist thesis clearly, though inklings of it go as far back as Leibniz.

The landmark work *Principia Mathematica* was published in three volumes starting in 1910. Next to Aristotle's *Organon* it is the most influential work on logic ever written. It consolidated earlier ideas about logic and number into an axiom system purged of all inconsistencies known at the time, and it rigorously deduced large portions of general mathematics from these axioms. Later, various flaws in it were addressed in alternative systems constructed by Ramsey (1931) and others.

The sense in which mathematics is reduced to logic in a system like that of *Principia Mathematica* was expressed by Russell with his usual lucidity as follows (1919, 194):

> So much of modern mathematical work is obviously on the border-line of logic, so much of modern logic is symbolic and formal, that the very close relationship of logic and mathematics has become obvious to every instructed student. The proof of this identity is, of course, a matter of detail; starting with premises which would be universally admitted to belong to logic and arriving by deduction at results which obviously belong to mathematics, we find that there is no point at which a sharp line can be drawn with logic to the left and mathematics to the right. If there are still those who do not admit the identity of logic and mathematics we may challenge them to indicate at what point, in the successive definitions and deductions of *Principia Mathematica*, they consider that logic ends and mathematics begins. It will then be obvious that any answer must be quite arbitrary.

Reading 'lower' and 'higher' for 'left' and 'right,' this passage can be taken as describing the continuum between the top two rungs of the ladder of reducibility.

To those who have never seen it done it may seem mysterious that mathematics could be wrung from logic alone. Where could the numbers come from? But numbers are closer to logic than might at first be imagined. Putnam has provided an example of how applied addition can be done using nothing but elementary logic (1975, 27). The example is an application of the arithmetic fact that two plus two equals four.

There are two apples on the desk.
There are two apples on the table.
The apples on the desk and table are all the ones in this room.
No apple is both on the desk and on the table.

Therefore, there are four apples in this room.

The whole argument can be formalized in the first order predicate logic with equality without actually naming any numbers as such. (The first premise, for example, translates into the formal equivalent of "There exists an x and there exists a y such that x is an apple and x is on the desk and y is an apple and y is on the desk and x does not equal y and for every z if z is an apple and z is on the desk then either z is equal to x or z is equal to y.") With the whole argument formalized in this fashion, the arithmetic conclusion follows by the laws of logic. This is not quite the way arithmetic is developed in *Principia Mathematica*, but it does illustrate the point that a logical capacity to deal with ordinary properties can be parlayed into an arithmetic capability.

Derivations of mathematics from logic are commonly accomplished by adding to first order logic a set of postulates for set theory, and then deriving mathematics within the set theory. Alternatively, the mathematics can be derived directly from a higher-order logic. In the latter case, essentially all the deductive power of set theory is built into the higher-order logic, so there is little substantive difference between the two approaches.

The various branches of mathematics are developed by introducing definitions and demonstrating properties of the defined entities on an *if-then* basis. For example, a theorem in group theory would have essentially the underlying form "If X satisfies the axioms of group theory then such-and-such can be said about X." Experience has indicated that nearly all of ordinary mathematics can be derived in this manner, though Gödel's incompleteness theorem imposes a theoretical limit on what can ultimately be derived from any particular set of axioms.

OBJECTIONS TO LOGICISM

The logicist thesis is by no means universally accepted among philosophers of mathematics. Numerous criticisms have been leveled against it (Black 1959), and they raise the question of whether it can be trusted

127

as a link in a reductive chain. But the criticisms do not necessarily apply to an evolutionary reduction. The issue that concerns the biological reductionist is this: Which if any of the objections raised against logicism carry over as objections to its role in the evolutionary Reducibility Thesis?

One of the earliest objections to logicism was the suspicion that the 'pure logic' to which mathematics can be reduced is not really so pure. Some of the logical postulates from which the logicist derivation of mathematics starts out are suspected of having empirical content. However, this objection to logicism is not troubling within the reductive framework, where all of logical theory is granted to be full of empirical content anyway.

Looking more closely, the logicist claim that "Mathematics is reducible to pure logic" is seen to fall into two separate parts:

Claim 1. Virtually all of ordinary mathematics can be derived from a few relatively simple axioms and rules of inference such as those of *Principia Mathematica*.

Claim 2. These axioms and rules of inference are purely logical in the sense of having no empirical content whatsoever.

Claim 1 is the technical aspect of the logicist thesis. It is based on actual formal derivations carried out in detail in *Principia Mathematica* and other works. The acceptability of Claim 2, on the other hand, hinges on philosophical judgments as to what 'purely logical' means. Evolutionary reductionism in mathematics requires only Claim 1. One could reject the traditional logicist thesis on grounds of doubting Claim 2 and still accept the evolutionary reduction of mathematics.

In the evolutionary reductionist scheme of things 'logical' versus 'nonlogical' is not a distinction that carries philosophical weight. It involves no great divide between *a priori* and *a posteriori*. How to use the words is more a matter of terminological convenience. The aspect of the logicist thesis that is just about arbitrary classifications – what truths shall be called 'logical' and what not – has little importance to logic as conceived in the evolutionary framework.

Logicism has also been accused of something called *if-then*ism. This doctrine and related forms of postulationism and deductivism hold that the proper subject matter of mathematics consists solely in the *if-then* connections between propositions. Especially important in this view are the deductive chains connecting the axioms that delineate a branch of mathematics to the theorems derived within the branch. These deductive chains are claimed to be the essence of mathematics. This philosophy is no longer widely accepted.

Whether or not *Principia Mathematica* is guilty of *if-then*ism, biological reductionism is not. It should be obvious that the biological reductive account imputes to mathematics far more than a concern with proof-theoretic *if-then* connections. The mathematics is part of a scientific ladder leading down ultimately to biology. Nothing could be farther from the spirit of reductionism than to limit the foundations of mathematics to the syntactic deductions on the top level.

One objection that has been raised against *if-then*ism is this (Resnik 1980, 131, 132):

> ... throughout the history of mathematics mathematicians have been convinced completely of results by means of arguments that fall short of deductive proof. ... These cases are particularly embarrassing to the deductivist [if-thenist], because he claims that a mathematical result always implicitly refers to axioms from which it is supposed to follow.

Such cases may be embarrassing to the *if-then*ist, but they are not embarrassing to a biological reductionist. As noted earlier, reductive logical theory does not demand that the axioms and formal inferences the metaanalyst happens to employ to describe fit organismic beliefs be accorded psychological reality within the organism. To the contrary, biological reductionism would be quite consistent with the hypothesis that mathematical knowledge is largely innate with no explicit axiomatic structure. A systematic scientific metadescription cataloging fit belief structures, though conveniently expressed axiomatically, should not be confused with the instincts themselves, or the actual processes by which the organisms may arrive at their conclusions.

It would be impractical to take up individually here all of the objections that have been raised against logicism. The question of which actually bear on the evolutionary reducibility hypothesis, and whether any are fatal to it, is open to exploration. If there are traditional objections to logicism that remain insurmountable even in the diminished form of logicism required for the evolutionary reducibility hypothesis, then the top rung of the ladder fails. But before settling for a decapitated ladder, each historic criticism should be reconsidered to see if it is still insuperable in the new biological context.

CHARGES OF CIRCULARITY

Now that the top rung of the ladder has been reached, there is a more comprehensive criticism to consider. A critic might complain, "There is something fishy here. General mathematics has only now been derived as the highest level of the ladder, yet mathematics was already in heavy use even at the lowest levels. Numbers and arithmetic were used in the averaging operations of the life-history strategy trees, for instance. What made it legitimate to use mathematics way back then, long before it had been derived at the top of the ladder? I sense a circularity, or possibly an infinite regress."

This sort of objection could be raised not only for mathematics but at all the logical levels. For example, deductive logic was already in heavy use in the treatment of the lower levels of the reductive chain all the way down to evolutionary theory, which as a science naturally involves deduction as much as any other science. What is to be made of this confusing situation?

The key to clarification lies (I think) in the distinction between *metalanguage* and *object language*. In a treatise written in French analyzing the structure of the Russian language, French would be considered the metalanguage and Russian the object language. There is nothing strange or objectionable about this. One can write in French about Russian, or in Russian about French. One can even write in Russian about Russian, provided the distinction between metalanguage and object language is carefully maintained.

In logic the term *language* is used broadly to include not only syntactical but also logical matters. In dealing with languages and their logical structures, modern investigations in formal logic and metamathematics have leaned heavily on the distinction between metalanguage and object language. The distinction was brought into prominence by Tarski (1956) following its introduction by earlier logicians. The system of logic being defined or analyzed is called the *object language*, while the *metalanguage* is the system of logic used to do the defining and analyzing.

Often the object language is formalized while the metalanguage is left as an informal mixture of, say, English plus logic and general mathematics. Frequently the metalanguage is much richer than the object language, and this is not considered suspicious; nor is it considered circular if powerful metalanguage devices are used in the construction of the object language before their object-language counterparts have been defined. Primitive stages of the object language can be constructed using the full power of the metalanguage. The metalanguage may contain the object language in whole or in part, may be similar to or different from it, more expressive or less, more completely formalized or less. None of these situations is regarded as objectionable so long as the separation between metalanguage and object language is clearly maintained.

In our reductionist reconstruction of logic, the metalanguage has included the usual informal mix of English, logic, and general mathematics, plus an admixture of evolutionary theory. The exposition has been in this metalanguage and the various reducibility theorems are really metatheorems. In keeping with logical custom the full power of the metalanguage has been employed freely in describing all levels of the object system from the ground up. Thus arithmetic was used in the metalanguage description of the life-history trees even though at that point in the development nothing had been ascertained about the object-level existence of arithmetic in organismic cognition. Their black-box behavior was described in an arithmetically equipped metalanguage, nothing more.

The conventional phrase 'object language' may not be entirely appropriate to a biological development because what we have been dealing with is not so much an object language as an object organism. The evolutionary approach shifts the focus of logic from languages to populations of organisms. It would stretch the notion of a 'language' even farther than has been customary in conventional logical theory. For this reason the parallel with the standard metamathematical paradigm is imperfect. It is perhaps complete enough, though, to shift some of the burden of proof onto those who claim something is amiss.

It is not enough simply to *suspect* a circularity. Critics should be asked to specify exactly what the circularity is and where it lies. Many, on first encountering the meta/object differentiation in metamathematics, have vaguely suspected a vicious circle only to see their suspicions evaporate as they become more accustomed to the distinction. Their confidence increases upon learning that its use has been well established for nearly a century with no apparent ill effects. And even if there is a circularity of some sort, not all circles are vicious. In discussing the circularities alleged to be inherent in empirically based epistemologies Vollmer defends them as "nonvicious, consistent, fertile, self-correcting feedback loops" (1987, 200).

131

Finally, to those who would reject the biological reasoning about logic on grounds of perceived circularities, it seems fair to pose the question of what sort of reasoning about logic they *would* accept as noncircular. Vague circularities are easily imagined for just about any theory of logic. To reason about logic, one has to have a logic. Is that a vicious circle? Or an infinite regress? If it is, we may as well give up all attempts to study logic rationally, and the ancients invented the subject in vain.

EVOLUTIONARILY STABLE LOGICS

If the classical family of logical systems can indeed be deduced from the Model 1 assumptions, this means that the classical logic as a whole is the maximally adaptive logic for Model 1 individuals. Those individuals behaving in conformity with the classical logic are selectively advantaged relative to those that conform to some other logic or to no logic at all. Any competing behavior patterns that conflict with the classical rules are evolutionarily unstable.

It will be convenient to have a term to describe this situation. Let us speak of the classical logic as the *evolutionarily stable logic* for Model 1. In particular, classical decision theory is the evolutionarily stable decision logic for Model 1, classical probability theory and the associated inductive inference systems are its evolutionarily stable inductive logic, the classical theory of deductive inference is its evolutionarily stable deductive logic, and (if the relevance of logicism is conceded) classical mathematics is its evolutionarily stable mathematics.

More generally, the evolutionarily stable logic for any model may be defined as the logic that is selectively advantaged, or optimal, for populations described by the model. The pressures of natural selection can be expected to tend to pull any population's behavior in the direction of its evolutionarily stable logic. Thus a Model 1 population is drawn in the direction of classical logic, and to the extent that natural selection gets its way its members will appear to reason classically.

REVIEW OF THE REDUCTION

Within the confines of Model 1 a way of reducing logic to evolutionary theory has now been outlined. How much confidence can be put in the reduction? This is largely a matter of how clearly each logical level can be seen to follow from the level below. Not all of the reductive reasoning has been formalized, but much of it has been made mathematically explicit here or elsewhere. The explicit parts, at least, fully satisfy Nagel's criterion that a reduction should ideally be established by exhibiting a formal derivation.

The phases of the reduction for which some degree of formalization is available here or elsewhere are these: (1) The tree-diagrams for life-history strategy theory were found identical in structure and solution algorithm to the tree diagrams of classical theory of decision under uncertainty; (2) A formal derivation of a leading axiomatization (Savage's) of classical decision theory from evolutionary stability theory has been provided (Theorem 4.1); (3) A mathematical derivation of subjective probability theory from decision theory is available in the work of Savage and others (Theorem 4.2); (4) The basic definitions are in place for a derivation of the pragmatics of deductive logic from this theory of probability (Definitions 5.1 and 5.2); (5) The semantics of deductive logic is derivable from the system of pragmatics (Theorems 5.1 and 5.2); (6) Well-known completeness results reduce the axiomatics of deductive logic to its semantics (Theorems 5.3 and 5.4); (7) Derivations of general mathematics from an axiomatically specified logic have been carried out by Frege, Russell, Whitehead, and other historic figures who have been concerned to establish the logicist thesis.

These points of formal confirmation are rehearsed here in case anyone should suppose the evolutionary reduction to be only a matter of casual speculation, like certain other putative reductions typically discussed more impressionistically than rigorously. The derivations show it to be more than mere speculation. It is intended to be taken literally as a real and probably fully formalizable theory reduction, with major portions already formalized. The fact that most of those who accomplished the latter formalizations did not have in mind a biological reduction does not diminish the support their work lends to such a reduction.

Those who would dismiss the Reducibility Thesis lightly should be asked what flaws they can point to in these various proofs; and if they cannot identify any, they should be asked what alternative explanation they can offer for this compelling lineup of formal derivations all pointing in the same direction. Appropriately assembled, the relevant derivations go a long way toward providing a direct formal proof of Nagel reducibility. Collectively they offer more formal support than is usually demanded of a scientific reduction. They constitute the hard core of the case for reducibility.

REASON RECAPITULATES EVOLUTION

As a looser way of conceiving the biological reduction of logic it may be helpful to think of a series of stages of abstraction from the particulars of the underlying evolutionary mechanisms. The different levels of logic then become successively distilled descriptions of the evolutionary process. Reason recapitulates evolution.

A stage-wise analysis of logic would start with evolutionary theory as a whole and abstract from it (i.e., ignore, drop) all biological considerations not included in life-history strategy theory. Next comes an abstraction from all nonbehavioral traits; what is left is decision and utility theory. Abstracting further from acts, consequences, and utilities leaves probability theory. Looking at the extreme event probabilities – those approaching one or zero nearly enough to be interpretable as belief or disbelief – one is able to define logical consequence and delineate the pragmatics of deductive logic. Abstracting from the belief-values of particular reasoners leads from pragmatics to semantics where connectives are construed as truth functions. Abstracting finally even from these, one arrives at axiomatic deductive logic, with general mathematics as its prolific definitional elaboration. Going back down the ladder, the abstraction operations are reversed to become reductive operations.

If reason does recapitulate evolution in this sense, the foundations of logic are already an integral part of evolutionary theory. The various levels of logic are, as it were, more or less detailed homomorphic images of the full evolutionary theory. Pursuing the essence of evolutionary stability for behavioral choice leads by stages to higher principles of logic and mathematics. Any crisp category distinction made between logical theory and the evolutionary theory that gives rise to it becomes arbitrary, for the various levels of logic are the successive residues of a stepwise abstraction operation. To extend Bertrand Russell's simile, we find that there is no point at which a sharp line can be drawn with biology to the left and logic to the right.

SUMMARY

The celebrated logicist thesis – that mathematics reduces to logic – takes the Model 1 analysis to the top of the reducibility ladder. At least, it is conjectured to do so if various historic objections to logicism can be overcome or shown irrelevant to the reductionist enterprise. The derivations of mathematical concepts from logical ones, constructed in support of logicism by brilliant minds, carry over to support the evolutionary reductionist position. Reducibility is transitive. If mathematics reduces to logic, and logic is reducible to evolutionary theory, then mathematics must itself be evolutionarily reducible.

What evolutionary reductionism requires in the way of logicism is in some ways more easily defended than the logicist thesis as originally conceived. The conventional distinction between 'logical' and 'empirical' knowledge, of which traditional logicism makes much, is not needed in an evolution-theoretic treatment of logic and mathematics. In evolutionary reductionism it is not the reducibility of mathematics to some arbitrarily conceived 'pure logic' that is of interest, but only its ultimate reducibility to evolutionary theory.

In the evolutionary literature, amazement is sometimes expressed at the fact that humans have evolved a capacity to discover and apply principles of abstract mathematics. The mathematical intuition is so impressive, and seems so far beyond the minimal survival needs of our ancestors! How could it have evolved? The reducibility arguments of this chapter may help lower the level of amazement somewhat. In their light, mathematical knowledge can be seen as an extension of internalized evolutionary processes – a ramification of a predictable instinct for avoiding structural instabilities in one's preference orderings. It doesn't have to be acquired separately somehow as an independent battery of Platonic ideas.

The summit of the ladder of reducibility has now been reached, for one especially simple population process model at least. The process in question, Model 1, imposes on the population members a certain fit logic, namely the classical logic. Though some of the reductive reasoning remains informal, there is a consilience in which each step of the derivation of logic becomes more plausible in the presence of the others. There is unity in the account as a whole.

8

Broadening the Evolutionary Foundation of Classical Logic

Model 1, the bundle of evolutionary conditions on which the ladder has rested to this point, is alarmingly simplistic. It is essentially a repository of all the simplifying assumptions that make for easy mathematical analysis of a population process. Among other restrictive features it does not allow for natural limitations on population size, nor does it take into account sexual reproduction.

In this chapter it will be seen that the latter two restrictions, at least, can be eased without changing the kind of logic produced. The analyses of how these constraints can be lifted are intended as case studies. They suggest that the biological foundations of the classical logic are not limited to Model 1, but can be made considerably more accommodating of realistic biological complexity. The evolutionary underpinning for the classical logic is a good deal broader than Model 1 alone might have made it appear.

TRAIT-NEUTRAL REGULATION

As noted by Darwin, a population growing unchecked at a constant proportional rate would soon run out of standing room. Exponential expansion simply cannot be kept up. It is important then, if realism is to be respected, that the effect of population-limiting contingencies be incorporated into the model somehow, and not ignored entirely as in Model 1.

The factors that constrain population growth are commonly called 'regulatory' conditions. *Regulation* will accordingly be used here as a cover term for all growth-limiting mechanisms. Regulation has to be taken into account in any process model that pretends to even rudimentary completeness over substantial evolutionary time periods. The interplay of growth and regulation is a critical aspect of all real population processes, and has been an important focus of the model-building activity of biologists and ecologists.

We shall confine attention here to regulatory phenomena that satisfy the following simplifying conditions with respect to the traits or strategies whose fitnesses are being compared:

(1) The regulation affects individuals with different traits equally; and
(2) The different traits contribute to the regulatory forces equally.

When both conditions hold we shall say the regulation is *trait-neutral* with respect to the characters in question.

Where there is trait-neutrality the regulative pressures do not bear more heavily on individuals with one of the traits of interest than on individuals with another, nor is heavier regulation brought about by the presence of individuals possessing one of the traits than by individuals possessing another. The traits and the regulation are independent of each other with no trait-differentiated associations in either direction. As a homely example, the trait differences might have to do with shade-seeking behavior in summer while the density-dependent regulatory limitations are due to scarcity of a certain food in winter. In such a situation it seems unlikely there would be any significant causal connections in either direction.

For many natural populations, the significant density-dependent causes of regulation are probably few in number compared to the myriad factors that can bestow differential growth on different characters. They are probably also independent of most of them. If so, trait-neutrality is in some sense the typical situation – the circumstance most likely to be encountered with respect to arbitrarily chosen decision problems. The factors that limit population size have long been a topic of debate, so the point is conjectural. But in any case trait-neutral regulation is the simplest case conceptually and makes a natural point of departure for mathematical analysis.

MODEL 2: A LOGISTIC PROCESS

It will be recalled that the underlying equation for the population process of Model 1 was

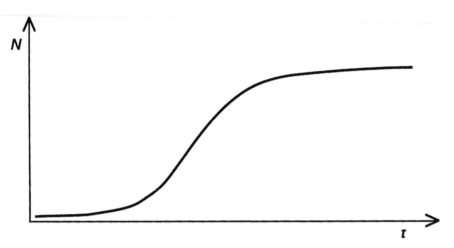

Figure 8.1. A logistic growth curve. Vertical axis represents population size, horizontal axis shows passage of time in seasons.

$$N(t+1) = RN(t)$$

where R is a constant finite rate of increase and $N(t)$ is the number of individuals in the population at the end of season t. To introduce a simple form of density-dependent regulation one could posit that there is a population-limiting effect proportional to the current population size. The equation then becomes

$$N(t+1) = RN(t)[1 - cN(t)] \qquad (8.1)$$

where c is a small positive constant representing the incremental downward regulatory pressure for each additional population member. The result of modifying Model 1 in this way will be referred to as Model 2.

The process has an equilibrium point K commonly described as the 'carrying capacity' of the environment. Reexpressing the equation in terms of K one obtains

$$N(t+1) = N(t) + (R-1)(1 - N(t)/K) \qquad (8.2)$$

This is one form of the well-known *logistic equation* (Lomnicki 1988, 11; May 1973). For values of R not much greater than 1.0 the growth of such a population exhibits the familiar logistic S-curve shown in Figure 8.1.

(For large values of R the population size changes become chaotic. Equation 8.2 is the discrete-case counterpart of the better-known logistic differential equation

$$dN/dt = rN(1 - N/K)$$

for the continuous case where r is the Malthusian parameter. An alternative logistic equation is sometimes encountered (Pielou 1977, 23ff.) which doesn't produce chaotic behavior but is otherwise qualitatively similar to Equation 8.2.)

Because regulation is now present the constant R no longer represents an actual population growth rate as it did in Model 1. Instead it is only one factor in the growth rate. In this office it is often called the *intrinsic* finite rate of increase. It is what the actual rate of increase would be if there were no density-dependent regulatory pressures.

Now suppose two traits, Trait 1 and Trait 2, are in competition in the population, and that the regulation is trait-neutral. Letting the intrinsic rates of increase of the two traits be R_1 and R_2, Equation 8.1 becomes the pair of equations

$$N_1(t+1) = R_1 N_1(t)[1 - cN(t)] \qquad (8.3a)$$

$$N_2(t+1) = R_2 N_2(t)[1 - cN(t)] \qquad (8.3b)$$

where $N_1(t)$ and $N_2(t)$ are the sizes of the Trait 1 and Trait 2 subpopulations in the tth season and $N(t) = N_1(t) + N_2(t)$ is the total population size. Condition (1) of the definition of trait-neutrality, requiring that the regulation affects the different traits equally, is satisfied because the density-dependent damping factor $[1 - c N(t)]$ is the same in both equations. Condition (2) – that the different traits influence the regulation equally – is also satisfied because $N(t)$ (and hence also $[1 - c N(t)]$) is a symmetric function of $N_1(t)$ and $N_2(t)$.

The growth curve of a population initially containing small equal numbers of the two traits is shown in Figure 8.2. Trait 1 has the larger intrinsic growth rate. It does better from the start and eventually becomes established while Trait 2 is driven down toward extinction. The total population size (topmost curve) has the usual logistic S-shape. In such a process the trait with the largest intrinsic rate of increase eventually wins out no matter what the starting proportions.

TREE PROCEDURE PRESERVED

In the life-history trees for Model 1, numbers assigned to nodes and branchtips represented actual growth rates. In Model 2, however, the growth rates change as the population size varies, so no single set of

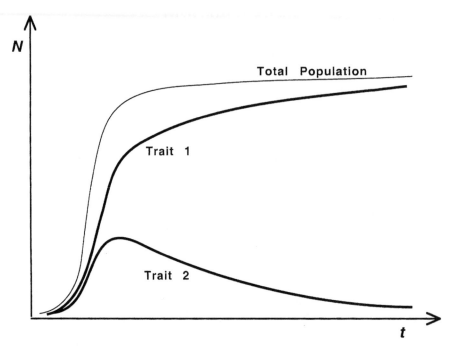

Figure 8.2. Two traits or strategies competing under logistic growth. ($R = 1.1$ for Trait 1, 1.09 for Trait 2, $c = 0.0001$.)

tree numbers for actual growth rates can represent conditions at all population densities. It would be inconvenient to have to draw different trees for different actual growth rates. The difficulty is easily circumvented, however, by redefining the tree numbers. In Model 2 they may be reinterpreted as intrinsic rates of increase instead of actual. Since intrinsic R-values remain constant in Model 2, this understanding results in a single tree that accurately reflects the population process under all density conditions.

This tree always indicates correctly the direction in which natural selection is working. Because the tree numbers no longer represent actual growth rates, absolute population flows can no longer be computed. Nevertheless the trait or character that the tree procedure picks out as fittest will indeed be advantaged. It is a trivial mathematical fact about life-history trees that when all branchtip and node values are multiplied by the same positive constant, the selective outcome of the tree procedure is unchanged. The same trait or strategy will be predicted to be the superior one as before the multiplication took place. So if in any season one were to multiply the tree numbers by the regulatory factor for that season to get actual growth rates, the choice of trait would be unaffected.

The upshot is that when the regulation is trait-neutral the bioanalyst can adopt intrinsic growth rates as fitness scores and not worry about adjustments for current population density. For intrinsic growth rates under trait-neutral logistic regulation, the tree procedure is preserved. We conclude that *the evolutionarily stable logic for Model 2 is the classical logic.*

MODEL 3: A GENERALIZATION

Logistic growth manifests only one type of regulation; there are many others. In general, trait-neutrality requires only that the regulatory function be of a kind that allows the process equations to be written in the form

$$N_1(t+1) = R_1 N_1(t) F(N(t)) \tag{8.4a}$$

$$N_2(t+1) = R_2 N_2(t) F(N(t)) \tag{8.4b}$$

for two competing traits. As in Equation 8.3 the parameters R_1 and R_2 are the constant intrinsic finite rates of increase of the trait-defined subpopulations, but now F (the regulatory function) can be any function of the total population size whose values are positive, bounded above by 1.0, and diminish with increasing population size. Let us call this set of conditions 'Model 3'. The logistic Model 2 then becomes a special case of Model 3.

Logistic models have been criticized on various grounds. One complaint is that the early upward-swooping portion of the standard logistic curve doesn't last long enough – that in most natural populations exponential growth actually continues unchecked until the equilibrium level is neared and the regulatory factors at last start to kick in (Wiegert 1974). The generalized Model 3 equations allow forms of growth that avoid this complaint, as well as other criticisms of the logistic model. Indeed since the restrictions on F are so minimal, it would be hard to think of any trait-neutral regulative schemes (stochastic, Markov, season-dependent, etc.) that are not covered by Equation 8.4 or minor elaborations of it.

Clearly, the earlier arguments for the preservation of the tree method under the logistic model carry over to all processes in the Model 3 family. Since the tree procedure is still valid in Model 3, all ensuing steps up the ladder of reducibility remain sound as before. We conclude that *the evolutionarily stable logic for Model 3 is the classical logic.*

MODEL 4: SEXUAL REPRODUCTION

Model 1 posited asexual reproduction or cloning. Few if any organisms with logical cognition actually reproduce in this way. A less artificial model, Model 4, is obtained if one posits instead a diploid, sexually reproducing population with Mendelian genetics, random mating, and a constant sex ratio. To avoid complications it may be assumed that the decision problems of interest are independent of gender in the sense that all decision-relevant events and consequences affect males and females alike.

In Model 4 a difficulty immediately arises concerning the interpretation of the life-history tree diagrams. If the values of R assigned to the nodes and branchtips are interpreted as average or expected numbers of offspring per population member as in Model 1, the value $R = 1$ will no longer represent a rate of growth in which the total population or subpopulation size stays the same. With a one-to-one sex ratio, there must be an average of two offspring per population member for the population size to remain constant; and if the sex ratio is anything other than one-to-one, the expected number of offspring a male must have will differ from the expected number a female must have. It would be inconvenient to have to work with separate tree diagrams for males and females.

These complications can be circumvented by adopting a slightly different interpretation of the tree numbers. Instead of denoting expected numbers of offspring, they can be taken to indicate expected numbers of *same-sex* offspring. For males only the male offspring are counted; for females only the female. When this is done the R-values valid for one sex will serve for the other also, and a value of $R = 1.0$ will again indicate a constant population size.

Another complication is that the population flow calculations become more elaborate when recombination is introduced. Suppose by way of illustration that two competing phenotypic traits are controlled from a single locus by a dominant and a recessive allele, with selection occurring by prereproductive mortality. Consider the tree of Figure 2.2 with the numbers now reinterpreted as expected numbers of same-sex offspring. The upper branch could be imagined to represent, for example, a phenotypic trait expressing genotypes that are either homozygotic for the dominant allele or heterozygotic, while the lower-branch character expresses the genotype that is homozygotic for the recessive allele. In such a situation the population changes at each succeeding generation can be calculated either from first principles or by applying standard formulae (Futuyma 1986, 156). One finds for instance that from an initial population in Hardy-Weinberg equilibrium containing the two traits in $1:1$ ratio, there is a change in one generation to a new ratio of $1.065:1$ in favor of the upper-branch trait. This is a somewhat slower rate of gain than was calculated in Chapter 2 for the asexual case. Nevertheless the gain is in the same direction as before. It is still the upper-branch trait that is selected for.

This illustrates a situation of considerable generality. Though the population flow computations are more involved in Model 4, they still support the selective choice indicated by the original tree procedure of Model 1. Recombination affects the rates of change but not their direction or the ultimate outcome as to which trait will eventually prevail. Thus the Model 1 tree procedure remains valid as a method of finding out which trait is selectively advantaged.

A similar situation would have obtained had the upper branch been taken to express the recessive allele. The upper-branch trait would still have won the evolutionary competition. Extended branching structures introduce no essentially new difficulties. The case of characters controlled from multiple loci is more complex but it is to be expected that the same basic situation will hold under rather inclusive conditions and the model can be extended to include these conditions accordingly. In all such cases the standard Model 1 tree-solution method, i.e., the classical decision-tree procedure, continues to give qualitatively correct results wherever all that is needed is a decision about which strategy is advantaged.

Because the original tree procedure still does its job of singling out strategies of maximal fitness, it can serve as in the earlier models as a basis for deriving classical logics. The tree diagrams still correctly compare the selective pressures on the phenotypes; and so long as the genetic system responds in the expected direction to these pressures by making the selectively favored phenotypes more plentiful, the classical logic will be supported. The conclusion to be drawn is that sexual reproduction under the conditions of the model does not disturb the reducibility argument. *The evolutionarily stable logic for Model 4 is the classical logic.*

A COMBINED MODEL

Models are easily constructed that take both regulation and sexual reproduction into account at the same time. For example, one could combine the defining conditions of Models 3 and 4.

For such a model the numbers in the tree diagrams would be interpreted as intrinsic expected numbers of same-sex offspring. The latter quantity can be taken as a measure of fitness in the model, as could any positive linear transformation of it. The tree procedure would accurately indicate selective advantage as in earlier models. Hence, for this model too the evolutionarily stable logic would be the classical family of logical calculi.

The combined model begins to approach what some biologists might consider a respectable starting model for some species. It is still simplistic, but not as hopelessly so as Model 1. The fact that this fairly ordinary model supports classical systems of reasoning encourages the hope that the classical logic, or something much like it, may have considerable biological validity over a wide range of evolutionary conditions.

LOGICALLY CLASSICAL MODELS

All of the foregoing models support a classical decision logic, a classical inductive logic, classical deductive logic, and (if logicism is accepted) classical mathematics. Let us refer to a model as *logically classical* just in case its evolutionarily stable logics are all classical in that sense. That is, in a logically classical model there is by definition selection for behavior consistent with classical reasoning at all levels of the ladder.

So far, attention has been confined exclusively to logically classical models. The question of which of the many known and possible population process models are logically classical remains to be explored. Each model must be examined individually, and nobody really knows

144

yet how common the property is. Not all processes are logically classical, however, as will be seen in the next chapter.

SUMMARY

How biologically robust is the classical logic? Though Model 1 is excessively simplistic, it is by no means the only population process that selects for classical reasoning. Some considerably more accommodating models are also logically classical. They include logistic and other models that allow for trait-neutral regulation of population size and simple Mendelian sexual reproduction. This suggests that the biological basis of the classical logic is secure within certain confines. The next step will be to explore what lies beyond those confines.

9

The Evolutionary Derivation
of Nonclassical Logics

There are respectable population models that are not logically classical. Such models give rise to evolutionarily stable logics, but the logics are nonstandard. The models are perfectly legitimate biologically, even though their logics involve departures from classical laws.

Three such models will be examined as case studies. Two of them (Models 5 and 6) have already appeared in the evolutionary literature in life-history tree form (Cooper 1981). One (Model 5) has also been analyzed in some detail in other papers (Cooper and Kaplan 1982; Kaplan and Cooper 1984), and has been reviewed in the context of related models (Godfrey-Smith 1996; Lomnicki 1988).

MODEL 5: A CHANGING ENVIRONMENT

Model 1 was a constant growth model. Each character-defined population of interest had a constant per-season rate of increase. Implicit in the constant growth supposition was the assumption that the environment does not change from season to season in ways that affect the growth rate. Such environments are said to be *temporally homogeneous*.

Dropping the temporal homogeneity restriction from Model 1 (but leaving everything else intact) produces a more general model, Model 5. In Model 5 the relevant environmental factors can change from season to season and the growth rate with them. Obviously the new model is capable of greater descriptive realism, for most real environments do fluctuate to at least some extent.

In Model 5 it will be assumed for simplicity that the environmental changes occur randomly through time. That is, knowing what the environment is like in one season is of no help in predicting what it will be like in the next. The different environmental conditions can have different long-run relative frequencies, however, with the different season types conforming to some definite probability distribution. For instance, in the simple case where there are only two different environmental conditions that can occur, we may picture Nature as having a certain weighted coin which she flips at the beginning of each season to determine which environment will be in force that season.

This coin should not be confused with the coins mentioned in the specification of Model 1. Recall that in Model 1 Nature had a weighted coin or die for each decision-relevant event that might happen to the individuals within a given season – e.g., a coin to flip for each individual to determine whether that individual will encounter a predator. The new coin or die for Model 5 is flipped or rolled only once at the beginning of each season to determine what the weighting of the within-season coins will be that season. Stated abstractly, the season-type probability distribution is a probability distribution over the intra-seasonal probability distributions.

By way of introduction to Model 5, consider the following decision problem. Suppose the question has arisen whether it would be more advantageous for the members of a certain population to build their nests in trees or on the ground. The environment fluctuates randomly with equal probability between two types of seasons. In one type of season the predators are sparse, in the other they are numerous. Nesting in trees offers some protection against the predators. It provides a break-even rate of increase of $R = 1.0$ in seasons of heavy predation. But tree nests are cramped and access to resources is restricted, so that even in seasons of light predation the growth rate is only $R = 2.0$. Nesting on the ground, on the other hand, makes individuals highly vulnerable to predation, so much so that in seasons of heavy predation the growth rate is really a shrinkage rate of $R = 0.5$. But in seasons of light predation the ground nesters do very well indeed with a growth rate of $R = 3.5$.

The nests must be constructed early in the season before there is any way for the organisms to detect whether it will be a season of light or heavy predation. Because of this there is no opportunity to make the nesting behavior dependent upon the current season type. The competition is between a simple inherited instinct for tree nesting and a simple inherited instinct for ground nesting. The problem data are

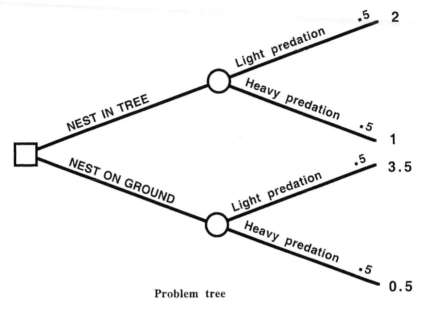

Figure 9.1. Is it is more advantageous to nest in a tree or on the ground when environmental conditions vary randomly between light and heavy predation?

* * * * *

diagramed in the tree of Figure 9.1. The reader might wish to try his or her hand at solving the tree before reading further.

Perhaps your approach was to follow the usual rules of decision theory, i.e., the life-history tree rules of the earlier chapters. Those rules call for maximizing the expectation by taking a probability-weighted average in each of the two main branches. The results are shown in Figure 9.2a. Such an approach seems natural enough because the maximization of expected utility is the foundation of classical decision theory, and has been seen to be reinterpretable in terms of fitness optimization. Biologically this procedure amounts to nothing more than maximizing the expected or average number of surviving offspring, a quantity that has been explicitly identified with fitness again and again in the evolutionary literature.

By that reasoning a population member maximizing its fitness would certainly have to nest on the ground. Indeed, Figure 9.2a shows ground nesting winning by the wide margin of 2.0 to 1.5. These figures could

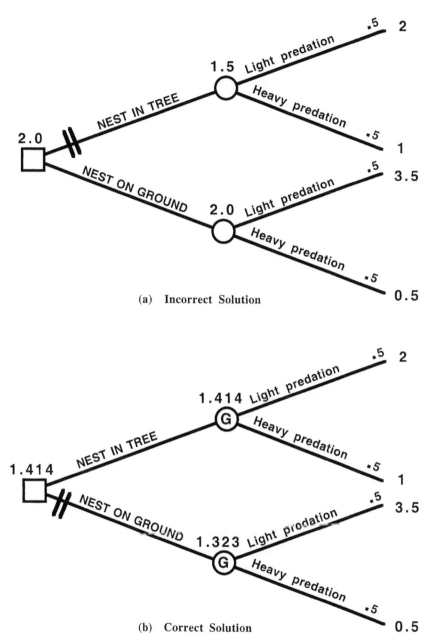

Figure 9.2. (a) The problem tree of Figure 9.1 solved incorrectly using the standard algorithm with arithmetic averaging. (b) The correct solution using geometric averaging.

mislead one into thinking that in an evolutionary race the ground nesters will win handily and take over the population in short order.

The trouble is, they won't. The tree nesters, not the ground nesters, have the advantage and will eventually win the evolutionary race. This can be understood when it is remembered that growth rates combine multiplicatively, not additively, over time. R is by definition the *multiplicative* factor by which a population's size increases in a season. Each season's ending population size is calculated by multiplying the previous season's ending population size by the growth rate for the present season. Now, in any long series of seasons generally about half will be seasons of light predation and half heavy, because Nature's season-type coin was stipulated as fair. Multiplying together a long series consisting of half 2s and half 1s yields a larger product than does an equally long series of half 3.5s and half 0.5s, as the reader can easily verify.

To calculate a long-run rate of growth, the general method is to take a product of the growth rates for all the different season types with each rate raised to a power equal to its probability of occurrence. Thus the tree nesters have a long-run growth rate per season of $(2)^{0.5} \times (1)^{0.5}$ = 1.414 while for the ground nesters the rate is only $(3.5)^{0.5} \times (0.5)^{0.5} =$ 1.323. In the short run either trait could be lucky enough to gain a temporary lead, but given sufficient time the tree nesters will overwhelm the ground nesters.

These computations amount to the calculation of a *geometric mean*. It is recognized in careful expositions of mathematical population biology that geometric means are what is called for when comparing growth rates averaged over time (Crow and Kimura 1970). If you forget and use the arithmetic mean you risk drawing wrong conclusions, as the example demonstrates.

Some naturalists like to explain the point in terms of variance (Gillespie 1974, 1977; Real 1980). A low variance among the possible rates of increase contributes to fitness. That is how tree nesting, with possible growth rates 2 and 1, manages to beat out ground nesting whose possible rates 3.5 and 0.5 exhibit more variance. Unlike the arithmetic mean, the geometric mean is sensitive to variance; lower variances tend to result in higher geometric means.

To make the tree method work in Model 5 it is necessary to take geometric instead of arithmetic means at all chance nodes associated with seasonal differences. The result is illustrated in Figure 9.2b, the correct solution to the problem. A 'G' has been written in the chance nodes to indicate that geometric means have been calculated there. In more complex problems a decision tree for a choice to be made in a fluctuating environment will usually involve both arithmetic-mean and geometric-mean nodes, so that some nodes will have Gs in them and some not. The original Model 1 then becomes the special case in which no G-nodes happen to be needed.

STRATEGY MIXING

Clearly the decision logic of Model 5 involves a departure from standard decision theory insofar as geometric averaging is required at some nodes. It is a surprising logic in that it goes against the conventional idea that fitness requires the expected number of offspring to be maximized. But it has a still more startling aspect that shows it to be qualitatively quite distinct from the classical theory.

In the tree-nesters versus ground-nesters example, the genetically determined heritable trait of building the nest on the ground was assumed to be in competition with another genetically determined heritable trait of building in a tree. However, there is a third possibility: Each individual could, as it were, flip a coin to decide what to do. What is genetically determined and heritable would then be not a specific nest-building behavior, but rather the propensity to let an internal chance mechanism guide the choice. Assuming the chance mechanism operates independently from individual to individual, in a large population this will practically assure that some individuals will adopt one behavior and some the other. For instance, if for some genotype the internal chance mechanisms are like fair coins, about 50% of the individuals of that genotype can be expected to nest on the ground and about 50% in a tree.

The coin involved here is still another type of coin, not to be confused with the coins and dice used by Nature to determine events or season types. This is an internal coin used by the organism to deal with the environment. When an agent makes a choice on the basis of chance, what takes place is known technically as *strategy mixing*. The internal chance devices used by the individuals might appropriately be described as their strategy-mixing coins or dice.

In Figure 9.3 the earlier example has been extended to show the coin-flipping strategy as a third possible character in competition with the other two. Each individual possessing the new trait, FLIP COIN, behaves as though governed by some internal chance device

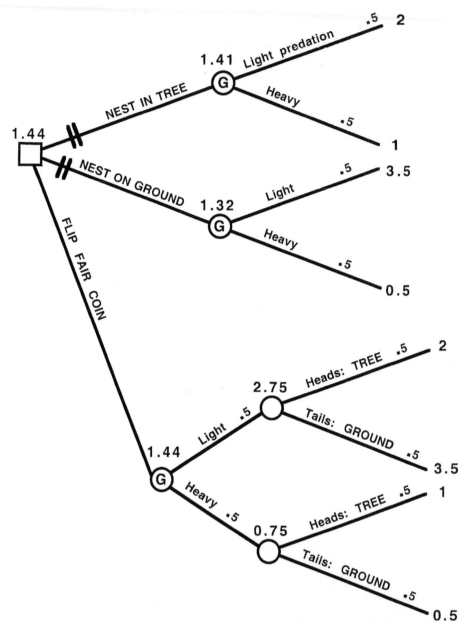

Figure 9.3. The tree of Figure 9.2b with an additional "coin-flipping" trait considered. The mixed strategy proves superior.

that gives it a probability 0.5 of building in a tree and 0.5 of building on the ground. When the geometric mean is calculated for the coin-flipping trait, it is found to have a long-run rate of increase of $R = (2.75)^{0.5} \times (0.75)^{0.5} = 1.44$. It is fitter than either of its pure-strategy competitors!

Strategy mixing will therefore be selected for. If natural selection has its way, what will eventually appear phenotypically is a stable mixture of tree-nesting and ground-nesting individuals. It pays in such a case for the population members to maintain amongst themselves this diversity of behavior, for if they don't they can be invaded by others that do.

The superiority of strategy mixing under these circumstances is perhaps unexpected. A strategy-mixing advantage is not ordinarily anticipated in classical decision theory. In game theory, as opposed to decision theory, it is widely known that appealing to chance can be a reasonable thing to do when each opponent is trying to second-guess the other. In Model 5 however there is no game-theoretic adversary, no opponent trying to second-guess the individual, only a neutral Nature creating a randomly changing environment. Nor is genetic maintenance of variation involved, for the strategy-mixing variation in question is intragenotypic. A single genotype produces a behaviorally polymorphic population by coding for coin-flipping. The possibility of adaptive strategy mixing in nongame-theoretic evolution was first pointed out by Levins (1962, 367–368) and the nesting example shows how it plays out on an intragenotypic level of analysis.

It seems entirely possible that a lot of coin-flipping is going on out there. It is hard to say what the specific chance mechanisms causing the strategy mixing might be. The phenomena that supply the internal coins or dice could include developmental noise (Waddington 1957, 39), random physiological variations, or sensory reception of random environmental cues. The theory predicts only that there will be some kind of intragenotypic strategy mixing.

WEIGHTING THE COIN

The example assumed the internal strategy-mixing coin to be fair. It is natural to ask whether an appropriately biased coin might be even more advantageous. That is indeed the case. When all members of the same genotype use the same genetically bestowed weighting an optimal weighting can evolve. If the population members were to flip coins weighted two to one in favor of tree nesting, i.e., to adopt strategy-mixing probabilities of 2/3 and 1/3, a slightly higher geometric mean rate of growth would be experienced. At that mixture the long-run geometric mean population growth rate is maximized.

A general formula for determining the optimal mixing probabilities is readily derivable (Cooper and Kaplan 1982, 143). Suppose there is random temporal variation between seasons of Type 1 and Type 2, these season types having probabilities of occurrence P_1 and P_2 respectively. Where it is advantageous to mix strategies between Strategy 1 and Strategy 2, it can be shown that population growth will be maximized if the strategy-mixing coin is weighted to give Strategy 1 the following probability of being chosen:

$$\frac{P_2 R_{2,1}}{R_{2,1} - R_{1,1}} + \frac{P_1 R_{2,2}}{R_{2,2} - R_{1,2}}$$

Here $R_{i,j}$ is the expected number of offspring for Strategy i in a season of Type j. The corresponding mixing probability for Strategy 2 is one minus this quantity. For the example these optimal mixing probabilities come out to 2/3 and 1/3, as previously mentioned.

It is to be expected that a population responding to selective pressures would tend to evolve toward the use of such an optimally weighted coin. In the example, one would expect to find about 2/3 of the population nesting in trees. The process is robust in the sense that the chance mechanism could temporarily stray considerably from the optimal mixing value and still provide some strategy-mixing advantage.

COIN-FLIPPING ALTRUISM

Evolutionary altruism has been variously characterized. According to one rough definition an *altruistic* trait is one which reduces the fitness of individuals that possess it while benefiting the population or species in which it occurs. "Altruistic traits are bad for the organism but good for the group." (Sober and Wilson 1994, 534). It is another surprising feature of Model 5 that it creates conditions conducive to a little-known form of altruism.

By way of illustration, consider a slightly modified tree-nesting example in which the concern is not with seasonally varying levels of predation but instead with high winds that arise unpredictably in about half the seasons. In seasons in which the winds are light, nesting in trees is more advantageous because that is where the fruit is. But tree nests are vulnerable to high winds. Ground-nesting is less advantageous in the absence of high winds, but high winds actually help the ground-nesters by providing them with windfall fruit. Plugging some hypothetical figures into this story we get the tree diagram of Figure 9.4.

The diagram shows an advantage to strategy mixing. Applying the formula, it can be calculated that the mixing advantage is at a maximum when the internal coins that the organisms flip are weighted to give odds of just under 7 to 3 in favor of tree-nesting. Coin-flipping at or near those odds causes the population to grow faster over long time periods than either pure strategy would, so it will be selected for. If selection is unimpeded the population will eventually settle into a stable distribution with about 70% nesting in trees and 30% on the ground.

The example is similar to the previous one. How remarkable, though, is the behavior of the ground-nesters in this case! By nesting on the ground they are opting for a fifty-fifty gamble between rewards of 3.0 and 0.5 when by tree-nesting they could just as easily have obtained a fifty-fifty gamble between the larger rewards of 4.0 and 1.0. The gamble they actually choose is clearly the poorer gamble for them as individuals.

It would be hard to think of any reasonable way of calculating fitness under which the decision to nest on the ground would have an immediate individual advantage. Certainly ground-nesting is not selfishly justifiable in terms of expected number of offspring for the individual. Yet, considering the salutary effect of their choice on long-run population growth, the ground-nesters cannot be said to have blundered or to be poorly adapted. The only plausible explanation is that they are sacrificing for the good of the population as a whole.

The ground-nesters could have had better prospects by opting deterministically for tree-nesting, but they flipped coins instead and got stuck with the personally less advantageous gamble. To accede to the collective coin-flipping regimen is bad for the individual but good for the group. Accordingly, such behavior might appropriately be called *coin-flipping altruism* (or more stuffily, *strategy-mixing altruism*). It is a subtle form of altruism in which, for the common good, the individual accepts a personally worse gamble even though a better one is available. Such altruism is robust and can be maintained within

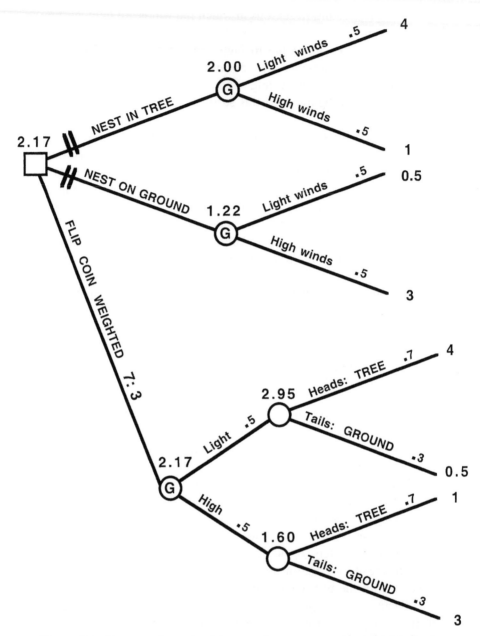

Figure 9.4. The superior mixed strategy bestows on some individuals the altruistic trait of ground-nesting.

a group indefinitely with no need to postulate competition with other groups.

A COGNITIVE SPECULATION

As mentioned, several mechanisms are possible which could in theory supply the needed strategy-mixing randomization. In the case of organisms with advanced cognitive capacities the possibilities multiply and are at present even more a matter of speculation. The coin-flipping could in principle take the form of a separate post-processing stage of cognition for randomizing choices, playing somewhat the role of a random number generator in a computer. Likelier, though, is a certain salutary looseness throughout all or part of the relevant cognitive activity, producing a degree of random unpredictability.

Such a favoring of randomization could have philosophically startling implications. To speculate a little concerning the species whose cognition is most familiar to us, what humans feel subjectively to be indistinctness in their logical thinking could conceivably be due in part to coin-flipping. When a decision or conclusion is a close mental call and there is vagueness involved, the vagueness could possibly be there partly to supply a potential strategy-mixing advantage. Thus the prospect arises of a sort of 'adaptive imprecision' in the ratiocinating activity of organisms with advanced cognition. An evolutionary epistemology is suggested that under some conditions justifies, or at least predicts, a small amount of mental sloppiness.

The very idea of adaptive imprecision flies in the face of traditional notions of rationality. We have all been taught by our mentors that crispness and exactness are to be prized in rational deliberation. Quantities are to be dealt with as accurately as possible and distinctions kept sharp and clear in the mind's eye. But from the coin-flipping theory we learn that rigid exactitude does not always promote maximally fit cogitation. Something looser and chancier can be fitter under some conditions. A little mental muddiness, a little blurring of category boundaries, a little use of approximation where exactness would be possible, could sometimes be more adaptive for the group than a keener form of reasoning.

A fundamental tenet of classical decision-theoretic reasoning is that of *stability*. According to the classical stability principle (not to be confused with evolutionary stability), if the same rational individual could be thrust repeatedly into exactly the same decision situation with access to the same knowledge on each occasion, the individual should manifest the same preference behavior each time. But where strategy mixing is adaptive as in Model 5, this classical principle fails to capture the full evolutionary situation. There, stability is not to be expected. Instead, a strategy-mixing instability will be rewarded. Not surprisingly, there is evidence that human preference judgments are in fact not especially stable (Fishhoff et al. 1982). Might the observed instability be explainable in part as mental coin-flipping?

Surely not all mental imprecision is adaptive. There would presumably be lots of thickheadedness in the world even if it were not for the coin-flipping advantage. But the theory suggests that selective forces which would otherwise work to eliminate cognitive imprecision are sometimes held in check to at least some extent by the benefits of intragenotypic strategy mixing. The extent is not yet known, nor is it yet known how many other conditions there may be beside temporal heterogeneity that can confer an advantage on coin flipping.

One is left with the unsettling conclusion that humans and other rational agents could at times be unwitting coin flippers. Though at this point it is little more than speculation, the theory hints at the possibility that we have all been dumbed down a little for the sake of our collective fitness.

THE NONCLASSICAL CHARACTER OF MODEL 5

The foregoing observations combine to support the conclusion that *the evolutionary stable logic of decision for Model 5 is nonclassical.* To recapitulate, a Model 5 life-history tree diagram requires non-standard calculations (geometric averaging). It does not follow the ordinary criterion of maximizing the expected number of offspring. Nor does it maximize expected utility, for the quantity maximized is not a standard expectation at all. There can be a strategy-mixing advantage. Coin-flipping altruism is possible. Classical stability is lacking. The model even suggests a possible epistemology of mental sloppiness. Now, the classical theory of decision was never like that. No standard axiomatization of it predicts such curious effects, at least not in any immediate or straightforward way. The behavior described by the model is in striking contrast with classical choice behavior in multiple respects.

The nonclassical character of the decision logic intrinsic to the Model 5 process is biologically primitive. It is not as though the organisms first developed some simpler logic such as the classical, and then somehow built the Model 5 logic up out of it. The predicted nonclassicality arises directly from the modeled process with no intermediate steps. We may reason *about* the organisms using a classical logic, but it is most natural to think of the organismic logic itself as nonclassical.

Someone might object, saying "What is so nonclassical about the new model? The tree calculations use standard addition, multiplication, exponentiation, and other classical mathematics. The biology may be more complicated than in Model 1 but it is still describable with garden-variety logical and mathematical machinery. Why speak as though a nonclassical element had been introduced?" But this line of thought confuses the resources of the metalanguage with the choice strategies of the object organism. To take the objection seriously would lead to absurdity because any rules of decision, however odd, could then be counted as classical so long as they could be described using classical mathematics.

Describing a logic in a classical metalanguage does not make it classical. This is an elementary precept that every logician will accept, but it has to be kept in mind. Suppose an English-speaking linguist were to set about analyzing an exotic language, hitherto unexplored. The linguist would naturally describe all the newly discovered properties of the language in English, at least at first. This does not mean that the exotic language is English. Nor does it mean that the speakers of the exotic language think like English speakers. It only means that it is expedient for the linguist to start out by describing the exotic language in a language familiar to the linguist.

The Model 5 logic is an exotic logic – a *biologic* if you will – that has yet to be explored systematically by logicians. It is the logic implicit in the behavior of evolutionarily stable Model 5 population members. It would be Procrustean to call that behavior, with all its nonclassical properties, classical. True, it is expedient at first for us to describe it using the formalism that is most familiar, the classical one. Having been trained to the classical logic we have little choice but to start out by adopting it as our metalanguage. This does not mean that the Model 5 logic is itself classical.

When the logic of a model such as Model 5 is described as logically nonclassical, what is meant is that the *organismic* logic defined by the modeled behavior is nonclassical. It must be nonclassical, for if it were classical what was the biological source of the classicism? Clearly not the likes of Models 1–4, for one is dealing here with Model 5. The classical logic enters only as a figment of the metalanguage. The organismic logic itself is defined by its own biology.

In Model 1 the underlying biology gave rise to trees of classical form, and it was provable (Theorem 4.1) that the resulting behavior follows classical laws of decision. In Model 5 that particular reductive train is broken. The reductive proof, which assumed classical trees, no longer goes through. The new model has its own trees, its own rules of choice which are biologically reducible but not classical. Examined as an organismic logic the classical decision logic has been shown to be biologically inadequate in a temporally heterogeneous environment. Thus it seems most natural simply to say that the Model 5 biology has given rise to a new, nonclassical, decision logic.

If Model 5's evolutionarily stable decision logic is indeed nonclassical, we have a case of a natural logic that describes adaptive behavior more accurately than the classical logic does. True, it is a fitter logic only in a randomly fluctuating environment; but since virtually all real environments are temporally heterogeneous to at least some extent, and the temporal change is usually effectively random to some extent, in principle just about all real population processes could involve similar nonclassical complications. The nonclassicality is ubiquitous.

MODEL 6: AGE STRUCTURE

It will be recalled that Model 1 assumed semelparity – that is, nonoverlapping seasonal generations with all offspring produced when the parent is exactly one season old. Lifting the semelparity restriction (but keeping the other Model 1 attributes) results in a more general model describing what are called *iteroparous* seasonal population processes. In this model individuals can produce offspring more than once per lifetime, say every spring. Let us call it Model 6.

Since the reproductive lifetimes of individuals of different generations can overlap, an iteroparous model is sometimes called an 'overlapping generations' model. The production of offspring is still seasonal and lock-step, but now each individual can produce offspring not only at the end of its first season, but also at the end of its second, third, fourth seasons and so on up to some maximum offspring-producing age of M seasons. Populations governed by such processes are said to be *age-structured*.

In illustration of the intricacies of age structure, suppose the members of a certain population are able to produce offspring at the end of their first and also at the end of their second season (i.e., $M = 2$). The two behavioral traits to be compared, let us again imagine, are nesting on the ground versus nesting in a tree. Predators prowl the ground. In a tree there is no danger from predators, but there is less room and resources. For a tree-nester, suppose the expected number of offspring is 1.0 in both its first and second season. For a ground-nester, suppose the expected number of offspring is 2.0 in both seasons provided the nest is not discovered by a predator. If a predator finds the ground nest in the first season, both parent and offspring perish so there will be no surviving offspring in either the first or the second season. If a predator finds the ground nest only in the second season, there will be 2.0 expected offspring from the first season and no offspring from the second. The probability that a ground nest will be found by a predator is 0.4 in each season. Which will win out, an instinct for ground-nesting or an instinct for tree-nesting?

$$* \quad * \quad * \quad * \quad *$$

As was done for Model 5, we present first a plausible but incorrect solution along classical lines, then the biologically correct solution. The incorrect solution follows blindly the classical notion that what must be maximized is an expectation, interpreted biologically as the expected number of offspring per individual. It also presumes that the decision tree should look classical. These preconceptions lead to the tree shown in Figure 9.5a. From it one can see, for instance, that nesting on the ground, if accompanied by the good luck of no predation in either season, results in four expected offspring, two from each season (third branchtip down). All computations are classical. The conclusion reached by this simple-minded analysis is that nesting in a tree is selectively advantaged.

Unfortunately that analysis is faulty. It does not take adequate account of the population's age structure. The offspring produced at the end of the second season cannot simply be added to those produced at the end of the first, because they do not contribute equally to the growth rate. Age structure complicates population growth considerably because the timing of the growth-producing events affects the overall rate of population growth.

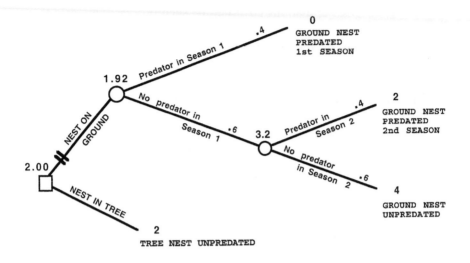

(a) Incorrect Solution

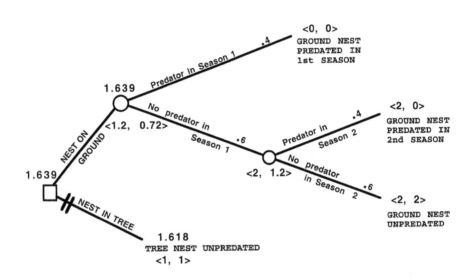

(b) Correct Solution

Figure 9.5. Two life-history strategy trees for an age-structured population. (a) A classical solution tree incorrectly predicts that tree-nesting is
advantaged. (b) A nonclassical tree maintaining age-specific distinctions in
fitness and applying Lotka's equation correctly predicts ground-nesting is
advantaged.

A well-known mathematical treatment of age-structured growth runs as follows. Let the expected number of offspring produced by each population member at the end of its first season be R_1, at the end of its second season be R_2, and so on through R_M. It is known that such a population will settle fairly quickly into a steady-state age distribution (except in pathological cases). After it does, its overall finite rate of increase will be the positive real root of the Lotka (or Euler-Lotka) equation

$$1.0 = R_1 x^{-1} + R_2 x^{-2} + R_3 x^{-3} + \ldots + R_M x^{-M}.$$

The theory underlying Lotka's equation will not be presented here but is readily available in the evolutionary literature (Mertz 1970 is especially lucid). In the special case of interest for the example, the case $M = 2$, the equation is easily solved to give a steady-state growth rate of

$$R = 0.5 R_1 + 0.5 \sqrt{R_1^2 + 4 R_2} \; .$$

In applying the theory it is to be noted that the possible consequences have, like the population itself, an age structure. Their values are M-length sequences of age-specific expected numbers of offspring. In our example the sequences are of form $< R_1, R_2 >$, representing the expected numbers of offspring produced at the age of one season and of two seasons respectively. The assignment of consequence values is

Consequence	Value
TREE NEST UNPREDATED	$<1, 1>$
GROUND UNPREDATED	$<2, 2>$
GROUND PREDATED IN 1st SEASON	$<0, 0>$
GROUND PREDATED IN 2nd SEASON	$<2, 0>$

An appropriate way of modifying the standard tree procedure can now be worked out. The corrected tree diagram for the problem is shown in Figure 9.5b. The consequences and their values appear at the branchtips as usual. Under each round node there is a two-number sequence representing age-specific expected numbers of offspring. These are calculated by working in from the branchtips from right to left, taking separate probability-weighted averages for each of the two possible reproductive ages. That is, the first member of each such pair is calculated by the usual probability-weighted averaging process from the first members of the right-neighboring pairs, and the second member is calculated similarly from their second members.

To complete the tree calculations it is necessary to reduce to a single fitness value each pair of numbers appearing under nodes immediately connected to the decision node. This is done by applying Lotka's equation as given above for the special case $M = 2$. The R-values so obtained are 1.639 for ground-nesting and 1.618 for tree-nesting. These values are written above the nodes and represent the stable growth rates of the two character-defined subpopulations. Since the ground-nesters have the larger fitness value they are selectively advantaged and will win out, contrary to the earlier analysis that forced the problem into a classical mold.

THE NONCLASSICAL CHARACTER OF MODEL 6

The modified Model 6 tree procedure is very different from the standard algorithm. Instead of just single numbers, number sequences (vectors) are assigned to nodes and branchtips. The computations involve a probability-weighted averaging of these sequences. The final step of solving Lotka's equation uses mathematics nothing like the elementary arithmetic that suffices for the standard tree procedure. Indeed, for larger values of M there may be no known analytic solution and successive approximations must be resorted to. Certainly the final fitnesses do not have the mathematical form of expectations of any ordinary kind.

One might look for tricks of alternative representation allowing a Lotka-modified Model 6 tree to be transformed into an equivalent one of classical form, as was done for Models 2–4. However, I have been unable to discover any such procedure. There probably is none because in an age-structured population it really does matter when things happen. The influence of a chance environmental event on population growth really is affected by the age of the affected individual. A Lotka-like complication would therefore seem inescapable in such decision problems.

Now, no standard axiomatization of decision and utility theory ever produced tree computations anything like those of the Lotka-outfitted trees. Thus it is hard to avoid the conclusion that *the evolutionarily stable logic of decision for Model 6 is not classical.* Model 6's decision logic is a different and more general logic than the classical. It subsumes the classical as the special case $M = 1$.

For this model too it might be objected that since all of the elementary tree computations are classical, nothing has occurred to upset the classical system. A Model 6 organism could get by using advanced classical mathematics, it might be argued. But again that line of thought is confused and loses the reductive thread. It forgets that the powerful classical mathematics defining the tree procedure is at this stage available only in the descriptive metalanguage. In an evolutionary development of the subject the object system resources have to be built up step-by-step in a biologically plausible sequence. When one comes to define the logic of decision biologically, one is still near the bottom of the ladder. The math from the top rung is not yet at hand, save in the metalanguage.

In assessing the significance of the model one cannot help reflecting that human populations are age-structured. Because humans are continuous breeders, not seasonal, their populational age structure is not quite so simple as that of Model 6. However, the Lotka theory generalizes in known ways to continuously breeding populations (Stearns 1992). It appears then that a more general model adequate to describe human age structure would introduce much the same sorts of complications as Model 6 does.

MODEL 7: A SMALL-POPULATION MODEL

Model 5 generalized Model 1 to the extent of permitting a temporally heterogeneous environment. We turn now to a more advanced population model – Model 7 – obtained by elaborating the model in three further directions. First, the assumption of a large population is removed. Secondly, there is assumed to be a "threshold of extinction" – some tiny population size (e.g., one member) below which no subpopulation can drop without disappearing. Thirdly, trait-neutral regulation of the kind already encountered in Models 2 and 3 is assumed.

For Model 7, the simple accommodation made in Model 5 for geometric averaging no longer suffices. In small populations the geometric mean growth rate is no longer the sole deciding factor in determining fitness because the danger of a random walk to extinction must also be considered. This is easily seen from any example in which two competing traits have exactly the same geometric mean growth rate, but for one trait the growth rate varies greatly with season type while for the other it remains almost constant. The two characters ought to be equally fit if the geometric mean growth rate is to be trusted as a fitness measure, but clearly they are not equally fit. The more variable character is much more likely to make an early exit in an unlucky random walk that happens to stray below the threshold of extinction.

What is needed for Model 7, then, is a more sensitive criterion of fitness. In Model 1 it was appropriate to measure fitness by the constant finite rate of increase. In Model 5 that measure was generalized to the geometric mean rate of increase to take account of a fluctuating environment. Now for Model 7 a still more general measure is needed to take into account the possibility of random walks to extinction in a small population.

The *expected time to extinction* (ETE) is an attractive candidate. It measures a trait's statistically expected persistence through evolutionary time, favoring those strategies that would tend on average to last longer. Originally proposed as a measure of fitness for small or island populations (MacArthur and Wilson 1967), it has received considerable mathematical attention. The mathematics of ETE can become complicated, but even without any mathematics it can be recognized as a plausible index of what ought to be measured. In Model 7 it would be possible to estimate the ETE of a character-defined subpopulation on a computer by repeatedly simulating its possible random walks, counting for each walk the number of elapsed seasons before the threshold of extinction is crossed, and averaging all the counts so obtained. The subpopulation with the highest average would be fittest according to the ETE criterion.

It would be desirable to have a tree procedure for Model 7, one that elaborates the tree procedure for Model 5 in such a way as to make it detect which character has the longest ETE. No such diagrammatic procedure is yet known. If one were worked out it would doubtless add further elements of nonclassical complexity to the Model 5 tree procedure.

The inference to be drawn is that *the evolutionarily stable logic of decision for Model 7 is nonclassical.* It is even less classical than that of Model 5. The classical decision theory must be generalized even beyond the changes required for Model 5 if it is to describe fit decision making in small populations.

ETE AS AN UNDERLYING FITNESS MEASURE

Model 1, though called a 'constant growth' model, does not exhibit absolutely constant growth. A Model 1 population is large but finite, and its growth is governed as though by environmental coin flips made independently for each population member. The collective environmental coin flipping will in general tend to produce slightly different growth rates in different seasons depending on the luck of the flips. In a large population the growth rate will tend to be *approximately* constant, thanks to the law of large numbers governing the coin flipping. The larger the population the less likely it becomes that there will be frequent significant deviations from the average rate, but there will be deviations.

The population size not only fluctuates, it fluctuates in a random walk. It is possible in principle for a strategy with a generally superior growth rate to have a run of bad luck and take a random walk to extinction before its competitors do. Hence average persistence becomes critical and in principle ETE is needed as the fitness measure even in Model 1. But it isn't used in practice, because in a sufficiently large population the probability of having such rotten luck is negligible. The approximately constant populational growth rates can be treated as though they were truly constant, and fitness comparisons made on that basis will ordinarily come out the same as if exact ETE computations had been made.

This line of thought suggests that in theory ETE is a more appropriate measure of fitness even for the likes of Model 1 than the 'constant' growth rate measure usually applied. It is theoretically more defensible though computationally less practical. The customary 'constant' growth rate measure supplies an expedient that gets around the computational complexities of ETE, yielding similar fitness comparisons with less trouble; but ETE is sounder in principle. Moreover, with a little thought it becomes apparent that similar remarks apply with minor variations to the other models that have been considered. Thus ETE could consistently be regarded as a more fundamental theoretical measure of fitness underlying all the models.

ETE AS A UTILITY MEASURE

If the foregoing argument has merit, ETE is really more universally relevant to fitness definition than has been generally recognized. I have argued elsewhere that it is general enough to subsume most other known persistence measures of fitness as special cases or approximations of some sort (Cooper 1984). ETE may well be the closest thing evolutionary theorists presently have to a fundamental fitness measure.

ETE lends itself especially well to evolutionary reasoning of the adaptationist variety. If 'fitness' is an intellectual construct defined to suit the purposes of the theorist, one might as well adopt the construct that best facilitates the reasoning most basic to the theory it is supposed to serve. ETE seems uniquely well-suited for adaptationist or optimization reasoning. All such reasoning is based on an underlying schema that runs something like this:

If X is fittest, then (other things being equal) X has the highest probability of being found to occur.

Here X is some trait being compared against other competing traits. What is wanted is a definition of fitness that will make this schema work well.

Of two traits X_1 and X_2, if X_1 has a longer expected time to extinction than X_2, then X_1 is likelier than X_2 to be found to exist if looked for at some randomly chosen moment in evolutionary time. Moreover, a trait's likelihood of being found extant when such a random time probe is made is directly proportional to its ETE. Indeed one would have to calculate the ETE to estimate that likelihood. Because of ETE's direct proportionality to a trait's probability of observed occurrence, it seems especially well suited to use in adaptationist prediction and explanation.

Defining fitness is not an easy problem in evolutionary theory. Under the reducibility hypothesis the significance of the problem extends into logic also, for it affects the interpretation of utility. In the reductionist theory, subjective utilities are subjective fitnesses. Once fitness is understood, utility is on the way to being understood too. How to measure fitness is therefore a logical as well as a biological question.

The foregoing ideas on fitness and utility might be summed up in the following tentative conclusions. To a first approximation a utility is measurable as an expected number of surviving offspring (or positive linear transformation thereof). This is a good approximation wherever logically classical population models are accurate enough. The natural unit of subjective fitness/utility – the 'utile' – is then the surviving offspring. To a somewhat better approximation, utility can be measured as a geometric mean growth rate. More generally still, fitness and utility are measurable as expected time to extinction. The utile then becomes the generation (or other time unit) and adaptationist reasoning is well served. It remains to be seen whether the study of more advanced process models will suggest still further refinements in how utility can be measured biologically.

It should be apparent by now that there are important population process models whose evolutionarily stable decision logics are non-classical. Such models yield the classical decision theory as special or limiting cases, but in their general form they are nonclassical. They have too much biological complication to select for behavior in the strict classical mold.

The question arises as to the character of the higher logical levels of the ladder in such models. No attempt will be made here to derive the evolutionarily stable inductive logic, deductive logic, or mathematics for any of Models 5–7. But even without doing so it is possible to contemplate what these systems of logic would be like if they were to be derived. The best guess is that they would probably be nonclassical, like the decision theory they build upon.

The conjecture here is that the nonclassical disturbances noted at the level of decision theory will be found to percolate up the reducibility ladder to affect the higher rungs. For population models that are logically nonclassical near the bottom of the ladder, the associated evolutionarily stable logics are likely to be nonclassical at all higher logical levels. To confirm the conjecture directly would involve carrying out an ambitious program of theoretical research; the higher-level nonstandard logics would have actually to be derived. But the conjecture is plausible even in the absence of such research because the reductive ladder has such a tight vertical structure. The character of each rung is informed by the rung below. For a complication present at one level *not* to affect the level above it would require a lucky cancellation – a transition that just happens to abstract away all the nonclassical lower-level detail.

In a strictly disciplined bottom-up reconstruction of logic, one moves from evolutionary theory through life-history strategy theory to decision theory and notes that for some models the corresponding decision trees no longer conform to classical postulates for decision logic. Theorem 4.1 assumed trees of classical form, but in the troublesome new models its proof no longer applies as it stands and would have to be generalized if it is to be kept true to the biology. The Savage postulates are no longer derivable as-is, and it can be anticipated that the theory of probability constructed from them will have to be doctored. The same applies, probably, up through deductive logic and perhaps higher. The quirks of realistic model-making, though they first show up innocently down at the life-history strategy rung, are likely to permeate the whole ladder of logic.

TOWARD A MORE REALISTIC LOGIC

To develop a realistic evolutionarily stable logic for humans (or any other cognitively advanced species) would be an ambitious project lying beyond the scope of the present essay. Still it is instructive to imagine what the task would require. It would involve deriving a composite model, the idea being to arrive at something close to the natural logic of the species. Models 1–4 or similar logically classical models could serve as a starting point, but it would be necessary to move beyond them to incorporate the properties of logically nonclassical models. For advanced realism all the generalizations would have to be combined into a single unified model. The composite model would simultaneously feature population regulation, sexual reproduction, age structuring, fluctuating environments, and so on.

Doubtless there are logically nonclassical population phenomena beyond those identified here that should be included in the composite model. It might well be found that there is no end to the many directions in which a composite model could be extended in the interests of greater descriptive realism. If so it would be necessary as a practical compromise to single out a few directions thought especially likely to have a significant effect on the resulting logic, and concentrate on generalizing in those directions.

The most adaptive logic for populations conforming to such a composite model would surely depart from the classical logic in multiple respects. It seems a safe conclusion then that *the evolutionarily stable logic for humans and animals with advanced logical cognition is nonclassical at all levels*.

DESCRIPTIVE LOGIC

Descriptive logic studies how organisms actually do reason, blunders and all. Many studies have been made of the descriptive logic of humans and other animals. Reviews of early work are provided by Edwards (1954, 1961) and Luce and Suppes (1965). More recent surveys are to be found periodically in the *Annual Review of Psychology*. The studies have ranged in methodology from word problems to putting subjects in choice situations and observing the choices made. Observations of interestingly nonclassical choice behavior include decision-theoretic anomalies such as the Allais Paradox. The latter takes the form of two pairs of gambles and evidence that the preferences of most experimental subjects between the gambles cannot be reconciled with classical decision theory (Allais 1953). Many divergences of behavior from the standards of classical inductive and deductive logic and mathematics have also been observed, those discovered by Tversky and Kahneman being the most widely known (Kahneman et al. 1982; Tversky and Kahneman 1992).

In many of these studies the classical logic is used as a basis for comparison. Sometimes the classical system is regarded as 'normative', in which case observed deviations from it are regarded as logical 'errors'. In other studies it is used simply as a convenient descriptive standard, with nothing said about its normative status. The thought in the latter case is that it is easier to codify observational results as deviations from some simple benchmark system than to start afresh.

I should like to raise here the question of whether the classical logic is really an appropriate benchmark. Suppose an advanced composite population model for humans were constructed along the lines suggested in the previous section, and its evolutionarily stable logic were worked out. This logic ought (arguably) to be a better reference point for descriptive purposes than the classical. Since it reflects actual selective forces, it could be expected to predict actual behavior more accurately. This is a testable hypothesis which we now offer:

POPULATION MODEL CONJECTURE: Human reasoning will be found to conform more closely to the evolutionarily stable logic of an advanced population model for humans than to the classical logic.

Testing the hypothesis would require working out an advanced human population model and its logic, or at least getting a good start on that formidable project. The testing would have to be done on logically naive subjects – bright but uncontaminated by training in formal logic. Perfect conformity of the subjects' behavior to the biologically derived logic is not to be expected because the model underlying the logic is bound to be incomplete, and the subjects' adaptation to the modeled process will naturally be imperfect anyway. Nevertheless it would not be unreasonable to expect the evolutionary logic to provide a better approximation than the classical to what is observed. Much classically anomalous behavior could turn out to be predictable in an evolutionarily stable system of logic tailored to the conditions under which the species actually evolved.

Evidence already collected using the classical logic as benchmark would be relevant to such a project. After the new descriptive logic has been defined, testing the population model conjecture would include seeing whether known subject deviations from the classical logic tend to deviate in a way that would have been predictable from the evolutionary model. If they do the Population Model Conjecture will have succeeded in making biological sense out of evidence already at hand.

SUMMARY

In this chapter, certain biologically legitimate population process models were found to be logically nonclassical. That is to say, they give rise to nonstandard decision logics – logics that resist formulation via the simple life-history and decision-tree procedures of the earlier chapters. They do not follow in any obvious way from classical axiomatizations, and some of them exhibit distinctly nonclassical characteristics such as the favoring of strategy mixing, altruism, and so on.

Though it has not been proved, their peculiarities probably propagate up the ladder from decision theory to complicate the higher logical levels. If so, biological complication in the underlying evolutionary models leads to nonclassical complications in the resulting logics. The standard logic is not necessarily universally fit at any level as it stands. It is a testable hypothesis (the 'Population Model Conjecture') that a nonstandard logic derived from a population model tailored to humans would yield more accurate predictions of actual human reasoning than the classical logic is able to do as ordinarily applied.

10

Radical Reductionism in Logic

By now it should be clear that biology has something to do with logic. But the Reducibility Thesis implies something stronger than that, namely, that biology is all there is to logic. The latter position I shall call *radical reductionism*. Radical reductionism in logic regards logic as completely reducible to biology with no leftovers. (Caution: The term *radical reductionism* refers here to radical evolutionary reductionism in logical theory. It is to be distinguished from the historical radical reductionism of Hume, Locke, and Carnap. Although there may be some common ground, the present theory departs sharply from earlier reductionist systems in regarding logic as itself reducible.)

Radical reductionism is an extreme position. There may be consistent philosophies of logic that are less extreme while still preserving some role for biology. The extreme philosophy is presented here with the idea that it is important to understand reductionism in its pure form before introducing compromises. It can then be decided more intelligently what compromises are called for, if any.

The claims of radical reductionism raise profound questions about what logic is and what systems of logic should be expected to accomplish. It might have been possible to ignore the hard questions if only the same laws of logic were obtained whether one started out from biology or from a more traditional starting point. But that is not the situation in which we find ourselves. It has become evident that some perfectly acceptable evolutionary models give rise to logics that are significantly nonclassical. The question of how these nonclassical logics are to be regarded brings matters to a head. It would appear that radical reductionism and classicism are in direct conflict (Cooper 1989).

SYNOPSIS OF RADICAL REDUCTIONISM

The principal tenets of the radical reductionist position are these:

(1) Evolutionary processes impose logics on populations in the sense that each population's governing process selects for behavioral conformity to the evolutionarily stable logic of the process.
(2) There is no other source of logic, nor is there any other criterion of logical truth or validity.
(3) In general, different population processes impose different logics on their population members.
(4) There is a special family of relatively simple population processes that give rise to the classical logic.
(5) Humans and other animals with advanced cognitive capacities are not accurately described by the population models of that family.
(6) Hence, the classical logic does not properly apply to human reasoning except as an approximation.

This position is radical in that it insists on reducing all logical concepts to their behavioral and biological roots, even if the results clash with traditional precepts. There is no attempt to preserve the classical status quo or compromise with standard ideas of what should be taken as normative.

THE RADICAL REDUCTIONIST

To understand radical reductionism in greater depth it is helpful to imagine how a radical reductionist might think about various matters. Such a thinker values synthesis and is uncomfortable with allegedly universal laws of no clear ancestry. The radical reductionist joins some other moderns in being suspicious of the traditional view that logical truth is utterly distinct from empirical truth. On a matter so basic as the nature of truth, would it not be more desirable to have a seamless account?

Logical validity is a touchstone. The radical reductionist defines logical validity in terms of fitness. Ratiocinated choice behavior is deemed logically sound just in case it follows strategies that are evolutionarily stable, i.e., not invadable by fitter strategies. For the

174

reductionist this criterion supersedes more traditional criteria of logical validity, and in cases of conflict is taken to be definitive.

The reductionist thinks it odd that anyone of the Darwinist persuasion could regard a logic as wholly correct that is adaptively inferior. A logic that loses its evolutionary races can hardly be counted sound in the overall evolutionary scheme of things, the reductionist thinks. Of what use would a logic be to an organism if it made it less fit than its competitors? In the reductionist's opinion, those who maintain that a logic can be valid without being evolutionarily stable have the burden of explaining why they would want to adopt a definition of logical validity that amounts to betting on a losing strategy.

Regarding classicism, the radical reductionist notes that in the most natural interpretation of the biological situation some population models give rise to behavior that is in accord with the classical logic and others to behavior that is nonclassical. Comparing the logically classical and logically nonclassical choice patterns, the reductionist sees no reason to regard the classical logic as special, preferred, or more 'normative' than any other evolutionarily stable logic based on a legitimate population model. Rather, each logic is valid with respect to the population process from which it derives. To the reductionist, the most appropriate logic is always the one derivable from a population model tailored to describe as fully as may be practicable the population of which the reasoner is a member. Such a logic will be valid for members of that population, or as close to valid as it is possible to come by formal methods.

Classical logicians emphasize the conceptual foundations of classical logic – truth-values, classical probability, and so forth. The radical reductionist accepts these precepts so far as they go but doubts that they add anything essential to what can be derived directly from a biological basis. Classicists admire the abstract generality of the classical foundations of logic. The reductionist does not admire the abstractness but is instead wary of it, being suspicious that it masks an inability to cope with biological complication.

The reductionist has his or her own evolutionarily derivable versions of the traditional theories. These not only supply everything needful, the reductionist feels, but also have the inestimable methodological virtue of being scientifically definable right down to the level of observation. The ultimate justification for logic shouldn't descend from high-level abstractions and assumptions, the reductionist holds. Rather, scientific justification comes up from below, from the

175

biological/behavioral foundations of which the traditional theories are only partial and imperfect reflections.

The radical reductionist sees the classicist as the victim of a needless dualism. A classicist, if biologically enlightened, may give some credence to the biological foundations of logic, since they seem to support it to some extent; yet will want at the same time to retain the traditional philosophical development of logic, as though it offered additional independent justification of its own for the classical laws. While conceding that the classical ideas have heuristic value, the reductionist thinks the classicist should wield Ockham's razor more fearlessly – should recognize that in the end the humble biological foundations are all that is needed.

To the reductionist, all scientific justification of logic flows upward from the bottom of the reducibility ladder. The reductionist believes it is possible to reconstruct the entire classical edifice of logic on that basis. But there is a catch. The reductionist knows that the classical logic can only be reconstructed biologically in straightforward manner if one starts out from a certain special class of models – the logically classical models. If one starts with other models one ends up with logics other than the classical. How then are these nonclassical logics to be regarded? The reductionist can see no reason why they should not be deemed as legitimate as the classical logic, normatively and in every other way. Indeed they may be more legitimate if they happen to be more descriptively realistic for the population whose reasoning is under consideration.

The reductionist is able to point to population models that give rise to nonclassical logics on at least the decision-theoretic level (e.g., Models 5–7). These evolutionary logics are regarded as raising difficulties for the classical decision logic. They call for special explanation if the classical outlook is to be maintained. And such special explanations appear to the reductionist to succeed only to the extent that they evade the question of where logical laws come from.

The reductionist concedes that there are as yet no logically nonclassical population models for which evolutionary laws of logic have been derived for rungs of the ladder above the decision-theoretic level. But it seems an easy surmise for the reductionist that when such a reconstruction has been accomplished, most such logics will differ, somewhat at least, from the classical logic on all rungs of the ladder. There is a potential array of reducible logical families, all legitimate in their place. The logician confronted with a particular application has the responsibility of choosing among the logics. The choice must be on the basis of which logic's underlying population model best describes the population of which the reasoning agent is a member, as well as on practical grounds of which models are simple enough to work with.

The classical logic is relatively simple, compared to logics derived from nonclassical population models. Some might regard its simplicity as one of its confirming virtues, since an elegant solution is often a correct one. The radical reductionist too admires simplicity, but understands that large biological systems are seldom simple. A simple mathematical model of an evolving population can hope to capture only a few of the most essential features of the population process. To include more of its features a more detailed model is required, and any number of further refinements are always possible. The successively more detailed models would allow more and more accurate logics for the population to be derived, but it seems likely to the reductionist that no simple logical formalism can ever guarantee perfectly fit ratiocination because no practically attainable model description for an organism with advanced logical cognition can ever be wholly accurate. As in biology, so in reductionist logic, the modeler has to make a compromise between realism and simplicity.

Approximation is all one ever really gets from formal logic, in this view. The radical reductionist believes it is important to recognize that limitation, especially insofar as it opposes the hoary tradition according to which absolute rigor and exactness is regarded as the only acceptable standard in logic. The classical logic's simplicity has the virtue of providing a convenient first approximation to any population logic yet known, and this makes it useful wherever a rough-and-ready approach will serve. There are even some indications that it is usually a rather *good* first approximation, though the matter begs for further exploration. But the reductionist thinks those who adopt the classical logic and apply it in the ordinary way should recognize that in choosing it they have already sacrificed some degree of precision.

LOGICAL ABSOLUTISM

The present-day theories that support classical systems of logic have inherited much of the logical absolutism of older logical traditions. The absolutist outlook has it that if a logic is valid at all it is valid, period. A sound logic is completely sound everywhere and for everyone, no exceptions! For absolutist logicians a logical truth is regarded as 'true in all possible worlds', making logical laws constant, timeless, and universal.

Logical absolutism is predisposed against reductionism in logic, or for that matter any other theory of rationality that questions the certainty of the one universal Logic. It sees logical laws as different in kind from scientific laws. Scientific laws are regarded as hypotheses dependent upon contingent fact, while principles of reason have a transcendental status. Logical verities are thought to be beyond challenge by ordinary empirical facts. They constrain the material world but remain unsullied by it.

Some absolutist attitudes toward logic have approached religious veneration. Seneca said, "Reason is nothing else but a portion of the divine spirit set in a human body" (Epistulae ad Lucilium, Epis lxvi). Thomas Aquinas wrote, "There is a certain eternal law, to wit, Reason, existing in the Mind of God and governing the whole universe" (Needham 1954, 538). Descartes put forward the interesting thought that clearly perceived principles of logic must be correct because God would not deceive us (Kline 1980, 183). Leibniz believed that God's options for designing the best of all possible worlds were constrained by logical laws expressible in a universal calculus: The Creator had free reign but only within logical limits. A twentieth-century commentator echoed the Leibnizian thought when he wrote ". . . [dictators] cannot alter the laws of logic, nor indeed can even God do so" (Ewing 1940, 217). Evidently, if it were to come to a contest between God and Logic, not all bets would be on The Almighty!

Asked where laws of logic come from, the absolutist is apt to reply that they are already known to the 'rational intuition'. The question of how they got into the rational intuition in the first place, or how it might be verified that the rational intuition is right about them, is not addressed. In effect such an absolutist is joining Seneca in declaring rationality to be simply "a portion of the divine spirit."

Some logicians, though hesitating to go so far as to deify logic, have nevertheless contrived in one way or another to invest logical principles with a comparable degree of imperviousness to empirical challenge. They maintain that rules of rationality need no justification, and indeed can never be justified without circularity, because they set the underlying standards on which any such justification would have to be based (Ayer 1956). Or in another argument, since logical theorems are universally valid by construction it would be useless to look for counterevidence. And so on. The upshot of all such absolutist views is a disinclination to look for the foundations of logic beyond logic itself. This unwillingness is wholly understandable, given the traditional way of preclassifying the universe of knowledge. After all, what could be more futile than to look for something prior to the *a priori*?

BIOLOGICAL RELATIVISM

The radical reductionist firmly rejects all such absolutism and replaces it with what might be called *biological relativism*. In the relativist view, each system of logic is dependent for its validity on the presence of some particular set of evolutionary laws describing the situation of the organisms to whose thought and behavior the logic is supposed to apply. Any formalized system of rationality is therefore understood to presuppose, explicitly or implicitly, some collection of biological conditions. A logic is scientifically valid to the extent that it accommodates the constraints imposed by the conditions. The laws of a logic are never absolute but always relative to the contingencies of some underlying evolutionary model.

Biological relativism of a general sort can be detected in the suggestion that intelligence might be something relative to the particular species or environmental complex under consideration. Being intelligent could conceivably involve something rather different for organisms that lived underwater, for example, or on another planet, than it does for humans (Ruse 1986b). If being logical is part of being intelligent, biological relativism in logic is a special case of the more general notion of biological relativity in all cognition.

Under the relativist conception, the one-logic-fits-all mentality has to be abandoned. Different population processes give rise to different logics, each logic valid for its own population. Philosophers have sometimes indulged in speculation about whether the laws of logic that are valid on earth would also be valid on an alien planet, perhaps a planet with a different physics. To the reductionist, the same logical laws aren't even uniformly valid everywhere here on earth, let alone throughout all possible universes. Different earthly species can have different logics.

Biological relativism is a less comfortable philosophy of logic than absolutism. With the validity of a logical system dependent upon the descriptive appropriateness of the evolutionary model from which it derives, as well as the chain of implications leading from the model to the logical laws, there is plenty of opportunity for oversimplification and error. In consequence a biological relativist has at most a certain degree of scientific confidence in any law of logic, never the 'logical certainty' traditionally ascribed to logical principles by absolutists. But to the relativist, logical certainty was always an illusion anyway.

THE DANGERS OF CLASSICISM

A classical absolutist is serenely confident of the essential correctness of the classical systems of reasoning. "I have been using the classical logic all my life and it has never led me astray," he or she will say. But how can he know? If the Population Model Conjecture is correct, the classicist has really been using something rather more sophisticated than the classical system in his daily affairs. On the rare occasions when he has actually constructed a complete chain of classical formal reasoning using pencil and paper, he may have checked his work ever so carefully *within* the classical framework. But has he ever tried to verify that the classical system itself is appropriate for him to be using as a human?

Questions such as this worry the radical reductionist, who has a darker view of the situation. The reductionist suspects that the classical logic, as ordinarily applied without regard to its evolutionary basis, is shot through with subtle biological errors and inadequacies. Fortunately the errors are usually slight enough so as not to cause much visible trouble, but they are there nonetheless and are hazardous. The reductionist position on this point is simple. It is that the classical systems of logic as typically practiced do not maximize fitness for humans because humans are not the product of logically classical population processes. The classical logic is unfit for human use, technically speaking.

To understand this radical stance it is helpful to study particular examples, such as the decision problem about nest-building that was posed in the last chapter by way of introduction to Model 5. The reader was invited there to solve that problem for himself or herself. Your first response was, perhaps, to apply the classical textbook decision tree algorithm to it in the obvious way. That would be natural enough because it has the look and feel of an ordinary decision problem. The writer admits to making the same mistake on first encountering the

Model 5 conditions, and suspects that anyone trained in classical decision theory would be liable to make the same error. The classical solution seems decision-theoretically appropriate, but on closer inspection turns out to be biologically inappropriate. It was a mistake, and it would have been a mistake (unfit) for the nest-builders to have acted out the oversimplified classical solution.

Now, if it is that easy to misapply the classical logic of decision even after attention has been drawn specifically to the complicating evolutionary conditions, how often does it get misapplied in ordinary decision problems when the biological background circumstances are not even recognized as relevant? The biology is always there, whether recognized or not, because the user of the calculus is always a biological organism. A conventional use of classical rules can easily produce unfit results without the user realizing it, especially if the user has been taught the classical logic and knows no other. The reductionist fears that such misuse of classical reasoning is ubiquitous.

It should not be imagined that the Model 5 complications arise only for creatures in the wild. They also bear on human affairs. Humans also live in an unpredictable temporally heterogeneous environment. Humans too face choices in which the available acts involve consequences with differing variances. It is we who are the nest-builders, the reductionist reminds us. For many human choice situations the decision trees ought properly to contain something comparable to G-nodes. That they do not is a biological flaw in the standard theory of decision as customarily applied.

True, Model 5 is much oversimplified where humans are concerned; but a more complete model for humans would include all the complications of Model 5 and more. If a simple-minded application of the classical decision procedure yielded a wrong result in the Model 5 nesting problem, is there any reason to think it could not produce wrong results in human choice-making? If an internalized classical decision tree would be unfit in other animals, what grounds are there for assuming it fit for humans? The radical reductionist sees no reason why humans should be regarded as exempt from these possibilities.

What Model 5 shows is that if any animal, human or otherwise, living in an unpredictable heterogeneous environment, were to apply the classical methodology in the biologically unsophisticated way taught in textbooks on decision theory, it risks suboptimizing its fitness. It would be acting out a biological miscalculation, a maladaptation, an error. To the reductionist this sort of mistake is a real, logical, mistake and fitness is diminished because of it.

Similar remarks apply to Model 6. A classical decision theorist, left to his or her own devices in solving the illustrative decision problem given for Model 6 in the last chapter, might well have constructed a classical decision tree and been satisfied it was correct. Unless prompted, the classicist would probably not even have thought to ask after the age structure of the population. But an analysis that fails to take age structure into account is a faulty analysis, as the example demonstrates. A classical decision tree would be an incorrect solution to the evolutionary problem. This is not a matter of subjective judgment, philosophical preference, or a mere difference of approach. The naive classical analysis is just plain bad biology and quite capable of yielding wrong predictions as to which strategies will win their evolutionary races.

The nest-building examples are counterexamples to the notion that the natural logic of animals such as humans is classical. Human populations are age structured. But what textbook on classical decision theory ever warned that the age structure of the reasoning agent's population must be taken into account? What course on classical utility theory ever taught Lotka's equation? The reductionist concedes that a naive classical analysis, applied blindly, might be lucky enough to yield good predictions much of the time. But that is because a classical solution usually tends to offer a good first guess, not because it is the correct analysis.

The situation is the same for Model 7 and other logically nonclassical models, as well as for advanced models that combine several logically nonclassical processes. A naive classical analysis, forging ahead with a biology-be-damned attitude, might fare better than random guesswork, but is by no means a full analysis revealing reliably the choice strategy that will actually be selected for. Humans or other animals who behaved in accord with the oversimplified classical analysis would have suboptimized their fitness. However, in light of the Population Model Conjecture it could be doubted that this is what they actually do when unencumbered by formal symbolic systems.

Higher rungs of the reducibility ladder are also affected. If inductive and deductive logic have their foundations in decision theory, as has been argued here and by others, biological inadequacies in the classical theory of decision can translate into failures of classical induction and deduction. Odd as it might sound, for all that is presently known *modus ponens* and other classical deductive schemata could well be biological oversimplifications, as could the standard rules of probability theory and other classical laws.

The difficulties on the higher levels are comparable to those for decision theory, except that specific biologically motivated rules for nonclassical deductive and inductive logics have yet to be worked out. Counterexamples to naive classicism comparable to those presented last chapter on the level of decision theory cannot yet be offered for deductive and inductive logic. But there is no reason in principle why deductive and inductive rules that are evolutionarily stable in logically nonclassical models could not be derived. Perhaps the derivation will eventually be attempted. There will then be nonclassical inductive, deductive, and conceivably even mathematical rules that are more reliable than the classical as principles of reasoning fit for humans.

An apologist for the classical logic might suggest that to get around the difficulties it is not necessary to abandon the classical logic, but only to learn to apply it in such a way as to take all the relevant evolutionary nuances into account. The reductionist agrees it would be a step in the right direction, but notes that a biologically adequate manual of application for the classical logic would be a formidable tome. The rules it contains would be vastly more complex than the classical logic itself. Working them out would involve constructing logically nonclassical evolutionary models and deriving their consequences in the form of rules of fit reasoning. If such a project were carried out in full, the reductionist suspects it would be more difficult than the task of developing a reductionistic system of logic to begin with.

REDUCTIONISM IN DEDUCTIVE LOGIC

To a classicist, the case for classicism might seem especially invincible in the case of deductive logic. A classical deduction allows a reasoner to move from true premises to a true conclusion. That is obviously a good thing to be able to do, the classicist opines. And you can rely on this truth-saving property of classical logic because the classical theory of deduction was specifically constructed to make it happen. There are propositions with truth-values, the connectives are defined truth-functionally, *modus ponens* is truth-preserving, and so on. Such thinking is bedrock classicism. A classical deduction *has* to be valid, by construction. The classicist even has an evolutionary gloss to put on the situation: Since it is presumably fitter to deduce true propositions rather than false ones, organisms who want to be selected for had better not violate the classical logic.

The radical reductionist is unconvinced. The reductionist suspects that the same fallacy flaws this argument as every other argument for universal classicism, namely, an unrecognized underlying assumption to the effect that all population models are logically classical. The argument for classical deduction could not be sustained if it were demanded that it be made biologically precise. The reductionist notes that it relies heavily on some rather abstruse notions, starting with the idea of propositions with truth-values. The classicist has the burden of explaining what such terms mean, else the whole enterprise of logic reduces to formalist games. Yet giving precise scientific meaning to metaterms like 'truth-value', 'proposition', and so on, is something classicists are usually reluctant to attempt.

The reductionist on the other hand can meet the challenge. The evolutionary derivations provide biological definitions of 'proposition' and 'true' (though not absolute truth). A proposition is an event and it is true for a Model 1 organism if and only if the organism believes it. This definition is not vacuous because it can be given scientific substance with biology-based definitions of proposition and degree of belief. The reductionist is pleased to offer such behavioral definitions for the classicist's use.

But the reductionist definitions are complication-free only for logically classical population models. For more refined models, the biological meanings generalize to something more involved, with details yet to be worked out. Thus the classical theory of deduction is in need of a general overhaul, in the reductionist estimation. As it stands it relies on suspiciously unreduced abstractions, and is far from transparently valid.

THE REDUCTIONIST CRITIQUE OF CLASSICISM

Any serious attempt to found a system of logic on evolutionary theory must proceed in a disciplined manner from the ground up. When attention is confined to hypothetical populations governed by logically classical models, the construction goes smoothly and produces the standard systems of logic. Higher levels of logic follow from lower and all turn out to support the classical laws. But as has been seen, cognitively well-endowed species are unlikely to be governed by such logically classical population processes. The nonclassical complications start at the life-history strategy level and produce a theory of preference relations that fails to support the reducibility theorems as given. From there the nonclassical behavior proceeds to affect the resulting decision theory and presumably all higher levels of logic as well.

The reductionist critique of the classical logic is that it ignores the latter complications. In the radical reductionist view, this neglect has made confirmed classical logicians into unwitting chauvinists for oversimplified evolutionary models. Impressive as the classical systems may be, they fail to reflect the fact that actual population processes are not as streamlined as classicism would implicitly have them. To the reductionist it appears that classicism has historically put too much faith in simplicity and elegance, and now the faith is turning out to be unfounded.

Traditional theories of logic would set up a fundamental asymmetry: Logic can be applied to biology but biology cannot be applied to logic. The reductionist, rejecting that asymmetry, applies biology to derive logic. In doing so the reductionist does not accuse the classical foundations of being entirely wrong-headed. It is granted that the classical systems are invaluable as a good first step, and are acceptable as far as they go. The problem is that they fail to encompass the whole biological reality. They make tacit biological simplifying assumptions that are rarely satisfied in reality. When the standard logic has been suitably biologized, its purely classical part emerges as a legitimate special case. The reductionist concern is that it is too special a case – too hypothetical and idealized – and what is most dangerous, not recognized as such.

The towering classical edifice totters on too narrow a biological platform, according to this critique. Classical logic as ordinarily applied simply fails to come to grips with the complexity of fit reasoning. As it stands it is too blunt an instrument. Until classicism is willing to look down at its own evolutionary roots it will remain simplistic and incomplete, quite capable of making wrong predictions and giving unfit advice.

SOME RAMIFICATIONS OF REDUCTIONISM

Radical reductionism notes that the decision logic implied by advanced population models is nonstandard, and foresees that its continuation into the higher levels of logic will probably turn out to be nonstandard too. Although nonstandard systems of logic and mathematics have been proposed before with some parochial success, still the prospect is unsettling from a classical viewpoint. The classical systems of logic and mathematics seem so secure and well tested, apparently have worked so well for so long, and are capable of predicting physical realities with such incredible accuracy, that to the classicist it is almost unthinkable they could be in need of serious modification.

Perhaps the incredulity is justified. But there are other possibilities to consider, if one were to permit oneself to speculate as a reductionist might. When an exact physical prediction is made and confirmed experimentally to a high degree of accuracy, both prediction and confirmation normally involve the use of classical logic. Thus as evidence of the perfection of the classical systems such a demonstration is somewhat circular. The confirmation could be corrupted by the same inappropriacies as the original calculation.

It is possible too that the practical differences that would ensue from using a nonstandard reductionist logic are so slight as not to matter in most ordinary circumstances. The classical logic might be quite good enough most of the time for most subject matter. If so it may only be under unusual conditions that the use of a special biologically derived logic would be justified.

There may be an analogy here with physics. Newtonian physics is quite adequate for most terrestrial purposes, but for problems involving speeds approaching that of light, relativistic adjustments are needed. Conceivably it is the same for absolutist classical logic as opposed to relativistic evolutionary logic. If so, as a practical matter the use of a biologically elaborated logic would be justified only for certain special classes of problems. Of course, an evolutionary metatheory of logic would be needed in order to recognize which those problems are.

Subtle inadequacies in the classical systems could easily have escaped notice. Because of the prevailing view that classical logic and mathematics are absolute, when something is found to be amiss in an observation or experiment scientists have tended to blame the experimental design, the measurement system, the interpretation of the logic, the way the logic was applied – anything but the system of logic itself. (There are exceptions such as quantum mechanics where the standard logic has been questioned.) Radical reductionism reopens the issue of whether the classical calculi really are as universally reliable and exquisitely precise as has been assumed.

In science a simple and intuitive conceptual theory is often thought attractive at first, but later found to require elaboration before it can be brought into conformity with testable reality. The reductionist sees the classical foundations of logic in this light. They are the product of centuries of careful deliberation, but it was top-down deliberation rather than a bottom-up scientific construction. Biologically the classical rules translate into plausible surmises but oversimplified ones. If and when a reductionist program of research into logic is pursued, it is foreseeable that the initial surmises will be found in need of elaboration and generalization as they are integrated into a scientific basis in biology.

CLASSICISM AND REDUCTIONISM CONTRASTED

To the classical logician steeped in convention, it is folly to think of the classical logic as having biological limitations. It would be a category error even to raise the question. The standard logic is thought correct quite independently of biology. "I have been given no reason to believe my system of logic is inadequate," the classicist will say. "The most that has been shown is that if I reason in the ordinary classical way, I risk reasoning unfitly. What if I don't care? What if I *am* a little unfit? That wouldn't make my reasoning faulty. Fitness is one thing, logic another."

To this the radical reductionist replies, "It is of course your privilege to think and act unfitly if you wish. But you should recognize that in attempting to reason at all, what you are really doing is obeying the impulse of your own evolutionary makeup. In effect you are trying to be your own consultant on population biology, your own life-history strategist. Fitness analysis is all there is to logic, ultimately. You have the choice of doing that analysis properly or improperly. It seems to me you have the burden of explaining why you would knowingly insist on doing it imperfectly when you have the opportunity to do it more perfectly."

This little exchange epitomizes the conflict between classicism and reductionism, between traditional and Darwinian attitudes toward logic. The issues involved are not easy ones. Reductionist logic carries out the idea of logic as fitness analysis. It attempts to provide a basis for describing actual logical behavior more accurately. In doing so it also offers fitter prescriptive rules to those who wish, when reasoning

187

consciously with formal aids, to be better life-history strategists in their own lives.

A confirmed classicist might defend any of a number of alternatives to radical reductionism. The simplest is the traditional classicism which presumes that logic and biology have nothing special to do with each other. To one who defends this view the reductive chain is explained away as a formal coincidence. The resemblance of decision tree diagrams to life-history diagrams is seen as pure accident. The reducibility theorems are ignored or dismissed as formalistic and empty. The logically nonclassical models are interpreted as evidence against the Reducibility Thesis, not against the classical logic. In this biology-precluding perception there simply is no biology of logic. The question of where the laws of logic come from is left unanswered or answered in some other way.

In a compromise view it is conceded that a fit logic has something to do with biology, but only on the decision-theoretic level. There it may contain nonclassical complications, but the higher logical levels are seen as self-sustaining on independent classical principles. The ladder of reducibility is broken in the middle.

There is also a live-and-let-live classicism that admits that logicians and biologists are fellow-travelers, but insists they have very different aims and goals. Such a classicist cannot understand how the reductionist can be so obtuse as to miss that simple point. Still another classical stance has it that the biologically derived nonclassical logics are really classical after all, in some extended sense of classicality. For a classicist who can rationalize that position there is no fundamental conflict between the classical and biological approaches and a peaceful coexistence can be declared.

To a traditionalist any of these positions could seem more acceptable than a thoroughgoing reductionism. Reductionism in logic is an unaccustomed way of thinking. To understand it requires a Copernican inversion of customary logical precepts – a nontrivial mental exercise. The classically trained mind resists entertaining the hypothesis that it is the biology that defines the logic. The reductive theory is easily misunderstood, easily dismissed on grounds of circularity, real or imagined. Reductionism would surely complicate logical metatheory and its application. It would call into question the universality of a standard calculus, one that is about as firmly entrenched as a theory can get. Conservatism in science is a legitimate and generally positive force, it can be argued, and one that would favor sticking with familiar logical precepts. Reductionism in any form is abominated in some quarters, where it is considered fair game to tar all reductions with the same brush. "Reductionism is a dirty word" to some, as Dawkins trenchantly remarked (1982, 118). "Reductionism is the traditional instrument of scientific analysis, but it is feared and resented" observes E. O. Wilson (1978, 13). All these potential objections militate against any easy or immediate acceptance of radical reductionism.

Yet none of them proves the reduction invalid. They argue its unpopularity, not its untruth. Perhaps a compromise approach could save some aspects of the standard orthodoxy; but it is not clear that any half-way measure can meet the reductionist critique fully, or offer the overall economy and consilience of the reductionist account. Further exploration is sorely needed to see what more is involved in a logicobiological unification.

As with any hypothesis, unanticipated developments could at any time weigh in against the reductionist theory. But in their absence it is hard to see why radical reductionism should not be taken seriously as a viable science of logic.

SUMMARY

The radical reductionist position is that evolutionary processes alone create logical laws, and different evolutionary conditions can create different logics. Each such logic, if properly deduced from an acceptable model of its defining process, is valid relative to the model in the sense of being evolutionarily stable with respect to it.

If this is accepted, no logic is uniquely and universally valid. In particular the classical family of logics holds no distinguished position over and above other evolutionarily derivable logics. It is correct relative to the elementary processes that give rise to it, but not beyond them. Moreover, the processes that lead to the classical logic are patently oversimplified. They are not the processes that actually govern humans or animals endowed with advanced cognitive capacities, but at best only approximations of them.

One is led by these considerations to the reductionist critique of the classical logic, namely, that for biological reasons the classical systems as ordinarily applied are not wholly reliable as guides to fit human and animal reasoning. More accurate systems await the derivation of evolutionarily stable logics for more realistic population models. Radical reductionism differs from classicism in admitting that the classical rules may be in need of such adjustment, especially in circumstances where the problem to be solved is sensitive to the biological background conditions. In the reductionist view the clarification of those circumstances should be a major goal of logical metatheory. Logic has to be made biologically accountable.

11

Toward a Unified Science of Reason

The outlines of an evolutionary theory of logic are now in place. In so brief an exploration of so large a topic it is inevitable that there are gaps to be filled and flaws to be fixed. But instead of filling in more details, it will perhaps be more useful at this point to ask how the theory, if accepted, would fit in with some other already established philosophies, and how it relates to certain bordering sciences.

LOGIC AS A SCIENCE

Traditionally, logic has not been regarded as a science – or at least, not an ordinary empirical science. Reductionism breaks with tradition by asserting it to be just that. It is true that logic deals with abstract intangibles such as propositions and probabilities, but since other sciences also have their unobservables and indirect observables this is not really a special distinction. Sciences are no less empirical for involving elements of a high degree of abstraction, provided the abstractions are rigorously defined from an ultimately empirical basis.

Logical theorems have traditionally been seen as sovereign principles for which no empirical confirmation is required. Their soundness is supposed to be observationally unassailable because they were never empirical laws to begin with. They are thought to be 'universally valid' and so not susceptible of observational disconfirmation. A caricature of the traditional view might run somewhat as follows. Logic was primordial. As a body of principles it has always existed and always will. After the physical universe was created and organisms appeared, their thought and behavior tended to adjust itself, to an approximation, to this preordained logic. A few complex organisms began to think rationally, meaning the organisms became especially well adapted to the preexisting constraints. The adaptation took place because the laws of pure logic are independently correct and it is fitter to reason correctly than incorrectly.

In this view, organisms were and still are being selected for obedience to fixed, universal logical norms. They are migrating toward absolute logical truth, a fixed star in a metaphysical heaven. Or in another metaphor, it is as if the organisms are being pressed by evolutionary forces into a logical mold, the mold itself having been shaped independently of any evolutionary process. There is supposed to be natural adaptation to the logical necessities just as there is natural adaptation to gravity or the geometry of space.

Evolutionary reductionism resists this view. It holds that laws of logic cannot be regarded as external constraints to be adapted to because there never were any separable, preexistent canons of logic to which to adapt. Logical principles are seen more as summary recapitulations of population processes than as eternal verities. They are adaptive not because they are metaphysical truths, but because it is adaptive for organisms to have an innately codified knowledge of the processes they are evolving under. The logical precepts induced by the processes are in no sense prior to the processes.

In the reductionist assessment, it is out of elemental evolutionary processes that the behavioral tendency to conform to logical patterns emerges. Failing to recognize these patterns as biological, philosophers have been pleased to call them 'analytic' rules, or principles of 'pure' logic, and have endowed them with transcendental properties. But once logic is seen as biological, the logical principles lose their transcendental character. A law of logic becomes a factual generalization in a scientific theory of a sort not fundamentally different from any other.

Thus logic emerges as a genuine science. To research logical principles is to research an ordinary science, albeit an especially ubiquitous one that permeates all the others. The proper scientific subject matter of logic is seen as a certain class of abstractions concerning predictable behavioral consequences of evolutionary phenomena. The consequences add up to what tradition has been pleased to call 'rationality'.

EMPIRICISM

Would such a science of reason be an 'empiricist' philosophy? Reductionism in logic is certainly not empiricism if what is meant by that term is the doctrine that the individual mind starts out as a blank slate, with no initial structure and nothing to work with but raw sense data. Reductionism is not sympathetic to John Stuart Mill's view that the laws of logic and mathematics are learned inductively from experience after birth. In the reductionist view the mind of a logically cognate individual has plenty of initial logical structure from the start, being the product of powerful selective pressures to reflect evolutionary processes. To be rational a mind needs enough innate structure to become its own population biologist – a tall order and far from the *tabula rasa* of Mill's empiricism.

If by 'empiricism' is meant instead the logical empiricism of Carnap and the Vienna Circle, reductionist logic is again nonempiricist. Logical empiricism of the Viennese variety relies on a sharp qualitative distinction between analytic and synthetic truths. This distinction is repudiated by evolutionary reductionism on the ground that it is short-circuited by the reduction. A distinction parallel to the traditional analytic/synthetic distinction can be made within a reductionist development of logic, but it refers more to different levels of generalization than to a philosophically portentous dichotomy between two worlds of knowledge.

If empiricism means ordinary scientific reliance on observation and empirically based theory-building, then evolutionary reductionism probably qualifies. An evolutionary science of reason would be empiricist in the sense that it regards none of the rich inherited mental structure needed for reasoning as informed by nonempirical truths. Rather, logical cognition is seen as the direct imprint of evolutionary processes.

THE *A PRIORI*

A related question is how a reductionist science of Reason would stand with respect to the existence of *a priori* truth. The reductionist stance is that there is no absolute *a priori* in logic, although to the individual organism it could seem as though there were. This would seem to be the position of Lorenz and Popper as quoted earlier. After denying the existence of absolute *a priori* truths, Popper recognized the existence of a 'psychological' or 'genetic' *a priori*. The latter could arguably be identified with what is genetically bestowed on the organism by the internal fitness-optimization apparatus that reductionism predicts.

In the radical reductionist version of philosophical history, the doctrine of an absolute *a priori* is seen as a hypothesis of desperation. Before evolutionary theory was available *a priori*ism was the only resort; it was unclear what else could be done. The elders looked upon Logic and said, "We do not know the father, so let us proclaim it a virgin birth." But now that a mature population biology is at hand and logic can be traced back to it, there is no more need to regard logical truth as *a priori*.

Historically, *a priori* necessity has at various times been attributed to theories other than pure logic. Examples include Euclidian geometry and Newtonian dynamics. History has not been kind to these ascriptions of *a priori*city. They have fallen one by one until only systems of logic and mathematics have remained (for some) as holdouts of faith in an absolute *a priori*. And if logical reductionism is taken seriously, that faith too is now challenged.

LANGUAGE GAMES AND FORMALISM

Some methodologists, especially those of logical empiricist leanings, conceive logical validity linguistically. They see logical truths as true by virtue of the rules of the language the truths are expressed in. It is the logician who makes up the syntax and semantics of a formal language, hence it is the logician who is in control of things (like Humpty Dumpty). What is logically true in the language must be true because its truth follows by virtue of the conventions and properties the logicians themselves have built into the language.

This philosophy is carried further by some who regard language as a kind of game. In their quasi-Wittgensteinian view the rules of a formal language are comparable to the made-up rules of game play. As such they can be agreed upon more or less arbitrarily. Naturally everybody has to stick by the same rules or the game won't work. In the language game known as classical logic everyone has to agree to stick by classical rules. Anyone who proposes other logics just isn't playing the classical game. It is not a matter of science or of an independent logical Truth, it is a matter of accepting established rules of play.

Those who hold with these ideas have a ready answer to the question of where the laws of logic come from. Their answer is that logical laws simply codify the properties of the language or language game that the logician invents. This is perhaps an answer of sorts, but a reductionist would regard it is a naive and dangerous one. The reductionist sees the rules of a logical language as heavily theory-laden. One cannot make them correct by fiat. As Putnam remarked, "If the rules of classical logic *were* really arbitrary linguistic stipulations (which I do not for a moment believe), then I have no idea how we are supposed to know that these stipulations are compatible with the aims of inquiry" (1969, 231).

To the reductionist, laws of logic have biological substance. They are not purely linguistic. Whatever may be the subject of the reasoning, evolutionary processes are always there in the background or there could be no rules of reasoning.

In its extreme form the linguistic perception of logic leads to a philosophy of logic and mathematics known as *formalism*. In the formalist program the emphasis is on symbol manipulation within axiomatic systems. The constraints on an axiomatic system are taken to be purely formal ones. Especially important is the requirement that no contradictions be derivable. Consistency is crucial because to an ardent formalist, formal consistency makes its own mathematical truth by guaranteeing the existence of model-theoretic interpretations. Formal analysis is the most reliable way to study that truth, the formalist thinks, because of the finitary character of symbol strings. The formalist attitude has been caricatured thus: "The only way to achieve absolute rigor is to abstract all meaning from the mathematical statements and to consider them for their own sake, as sequences of signs, neglecting the sense they may convey. Then it is possible to formulate definite rules for deducing new statements from those already known and to avoid the uncertainty resulting from the ambiguity of language" (Heyting 1956, 57).

Reductionism does not necessarily oppose all elements of the formalist program, but it emphatically rejects the notion that a consistent formalism makes its own logical truth. A formalist in the grip of such a belief is at the mercy of the formalism itself and blind to any fundamental shortcomings it may have. Logical principles are always heavily invested with biological meaning. Though it is possible to make up clever symbolic games and agree to play them, they are no more than that unless they connect somehow with the observable world.

Reductionism emphasizes the need to follow a logical theory down to its observational roots to see if it describes what it is supposed to describe in the way of fit behavior.

The sophisticated formalism of David Hilbert and other early twentieth-century metamathematicians is considered by some historians to have foundered on the rock of the Gödel incompleteness results. That however is not the important objection to it in the reductionist estimate. To a reductionist the trouble with Hilbert's formalism is simply its formalism, its inattention to substance. That has been the complaint about it by many critics from Frege on. Evolutionary reductionists now join in the old complaint, adding that the substance that is lacking is ultimately biological.

The chief rival of Hilbertian formalism was a school of mathematical thought known as *intuitionism*. Intuitionists such as Kronecker and Brouwer believed that all mathematical and logical analysis should take as its starting point the mind's firm intuitive grasp of the natural numbers. This led them to question certain classical laws of logic such as the law of the excluded middle. Because the intuitionistic point of departure is already far up the ladder, taking for granted concepts that are themselves in need of explanation from a reductionist perspective, it should be immediately apparent that intuitionism and biological reductionism are not especially compatible. Apart from a shared distrust of formalism and the classical logic they have little in common.

GRAND REDUCIBILITY HIERARCHIES

If the Reducibility Thesis has it right, the ladder of reducibility leads down at least as far as evolutionary biology. Might it extend down still further? Is evolutionary theory itself reducible to something more basic?

The Medawars (Medawar and Medawar 1977, 164) suggest that the principal sciences fall naturally into the following order:

Ecology/Sociology
Organismic Biology
Chemistry
Physics

They remark that there is something curiously compelling about this list. They note that any thoughtful scientist can see that the topics have been set in the right order, and that if the sequence were mixed up most scientists would be disturbed and want to put it back into the order shown. The reason for this, they say, is that "each science in the hierarchical table is in some sense a special case of the one below it and that any statement which is 'true' and 'makes sense' in any one of these sciences is also true in and makes sense in any science above it." Explaining that the relationships involved are essentially reducibility relationships, they comment that "reductive analysis is the most successful explanatory technique that has ever been used in science."

In an earlier but methodologically more sophisticated account, Oppenheim and Putnam proposed a reducibility hierarchy that is remarkably consonant with the Medawars' (1958). Their reductive levels are:

Social Groups
(Multicellular) Living Things
Cells
Molecules
Atoms
Elementary Particles

On each level the theory of the entities in question is reducible to the theory of the entities below, they assert; or at least it is 'microreducible' in the sense that various aspects or subtheories of the level above are reducible to aspects of the one below. The reductive levels suggest, they say, a "natural order of sciences." Physics is concerned mainly with the bottom three levels, biology mainly with the fourth and fifth levels, and so on. Evolutionary biology draws on the upper levels. The authors offer their hierarchy as a working hypothesis for the unity of science.

Such orderings are examples of what might be called grand reducibility hierarchies. They are intrepid attempts to discern the structure of science as an integrated whole. Without attempting to pass judgment on their merits, it is worth noting that if there is such a thing as a grand reducibility hierarchy of science constructible along some such lines, a reduction of logic to evolutionary theory would make it yet grander.

The formal sciences have traditionally been separated off from the empirical sciences and so excluded from the grand hierarchy. But if the biological reducibility of logic is accepted this need not be so, for the formal sciences are then empirical and belong with the rest. Our

197

reducibility ladder for logic would stand atop ladders such as the afore-described. The formal sciences could then be compared to the upper section of an extension ladder. The combined height of the original ladder plus the extension would be impressive indeed.

Building up from, say, the Medawars' hierarchy the combined ladder would look something like this:

Mathematics
Deductive Logic
Inductive Logic
Decision Theory/Game Theory
Ecology/Sociology/Population Biology
Organismic Biology
Chemistry
Physics

Is such a comprehensive reducibility hierarchy – one that embraces the logical as well as the empirical sciences – defensible? It is a stimulating idea but one that lies beyond the scope of the present exploration. We content ourselves here with raising the question.

ALLIED DISCIPLINES

How might a biological reduction of logic relate to other disciplines? The *cognitive sciences* seems especially closely implicated. Since the reduced logical theory has a behavioral, externally determined character one naturally looks to cognitive psychology, neuroscience, and allied studies to fill the reductionist's black box with plausible internal mechanisms. A biologically based science of reason would extend naturally in these directions.

Reductionist logic provides the basic insight into what the apparatus of logical cognition is for; namely, to do population biology. But it remains to say just how the activity is accomplished and what the enabling mechanisms are. How is a belief state structured psychologically or neurologically? Which utility values are inherited, which estimated once and stored, and which estimated on the fly as needed? By what process is sensory information organized and used to update subjective probability values? When one reflects that all knowledge, insofar as it is needed to guide behavior, ought (in the reductionist view) to be structurable in terms of acts, consequences, utilities, and belief states – or possibly other entities that can effect an equivalent ratiocination more efficiently – the cognitive challenge of discovering how these are represented and manipulated is glimpsed.

Linguistics too is a neighbor to reductive logic, and perhaps a nearer neighbor than might at first appear. The reductive theory of logic in the bare-bones form presented here is language independent with nothing assumed about the object organism's means of communication. Nevertheless a reductive logic has ramifications for a comprehensive evolutionary theory of language. An example was noted in passing in the matter of the indicative English *if-then* connective, for which logico-linguistic evidence has actually been collected. If the English *if-then* is indeed a pragmatic conditional probability connective as the evidence indicates, it is hard to see how its true meaning could be explained except in a logical context framed along the indicated belief-theoretic lines. That example could be a hint that linguistic capability is tightly bound up with logical cognition.

Reductively speaking, linguistic communication is analyzable in large part in terms of changes of belief states and utility estimates. Languages provide names for propositions or events, and linguistic utterances modify and compound these names in various ways to bring about fit belief state changes. Linguistic intercourse becomes a means of effecting adaptive changes in the subjective probabilities and utilities of the interlocutors. It can evolve, can be dependent upon population dynamics, and can require altruism in the evolutionary sense (Skyrms 1996). These and other considerations suggest the global framework for a comprehensive theory of language should be logical and biologically reducible.

In *artificial intelligence* (AI) there is a place for integrated theories of rationality inclusive enough to extend to coherent decision making, deduction, and induction, with linguistic and mathematical extensions an additional desideratum. Reductionist logic could offer such a framework. It is one that arguably captures much of Nature's infrastructure for intelligence. In attempting to apply such a theory, AI would share with cognitive science the challenge of finding economical and efficient internal representations of belief states and utilities. The logical theory alone does not supply these (at least not beyond toy domains) but it does offer a foundation upon which a more complete superstructure might be built.

Allied to artificial intelligence is the more recently conceived field of *artificial life*. A major activity in artificial life research has been the study of populations of programlike entities or data structures that are made to 'evolve' through many generations in a computer-simulated environment. Forces of natural selection and mutation are included as part of the simulation. The hope is that the individuals in the population will evolve under the selective pressures in the direction of interestingly adaptive behaviors evocative of 'life'.

In a sufficiently comprehensive evolutionary computer experiment – one in which the population members' logic is not preprogrammed but allowed to evolve – the logical behavior that evolves may not be classical. Indeed it will surely be nonclassical if the simulated population process happens to be logically nonclassical. The computer-simulated individuals will not end up as perfect Bayesian inducers or classical deducers, but will instead maximize fitness/utility using a more complex nonstandard reasoning that is fitter. Reductionist logic gives reassurance that there is nothing strange or disturbing about this. To the contrary, logicians might do well to look to artificial life research for ways of simulating complex population processes to see what sorts of nonstandard systems of logic might evolve.

Touching on all these overlapping areas is a branch of philosophy that threatens to become a science. *Evolutionary epistemology* attempts to investigate what we can know and how we can know it in light of evolution. Just as traditional epistemology has always included the study of conventional logic, so evolutionary epistemology should presumably include the study of reductionist logic. Should the reductionist epistemology of logic supersede more orthodox epistemologies? Logic has never been as settled and tranquil a subject as popularly imagined, and it could well become even less so as such questions are contemplated.

CONCLUDING REMARKS

Evolutionary reductionism departs radically from conventional philosophies of logic. It proposes that logic has biological foundations; that it is not only metascientific but scientific; not purely analytic but empirical; not *a priori* though it could seem so to the individual; not valid absolutely but only relative to particular population processes; not primitive but derivative; not simple but biologically complex; not exact but only approximate; not tautologously true but scientifically confirmable; never certain but at most highly probable; not ideal truth but factual generalization; not properly a branch of philosophy so much as of biology. It calls for a Copernican reorientation toward a more biocentric model of reason and the scientific method.

Whether these wholesale differences are counted as virtues or defects will depend on how well it is understood what the theory is trying to accomplish. The reductive account is able to give a straightforward scientific answer to the question of where the laws of logic come from. It is potentially capable of repairing subtle unfitnesses in classical reasoning. It would unite the two worlds of truth, the logical and the empirical. If it succeeds nothing will have been lost but a false simplicity.

The biological turn in logic could further the unity of the sciences by connecting the pieces of the logical puzzle not only to each other but to science in general. It could weave into a seamless whole various strands of ideas conceived by many eminent thinkers including evolutionists, logicians, psychologists, economists, and philosophers. How well it might ultimately succeed should not be judged solely by the modest prolegomenon ventured in these pages. Much remains to be explored, there are difficulties to be overcome, but the unity beckons.

> The philosophical mind unites where the pedant parts, he is convinced that in the provinces of both the intellect and the senses all things are linked together, and in his desire for synthesis he cannot content himself with fragments.
>
> – Johann Friedrich von Schiller (1789)

Appendix

Formal Theory

This appendix provides a proof of Theorem 4.1. More generally, it offers some minimal mathematical foundations for an evolutionary derivation of classical decision and utility theory, subjective probability theory, and deductive logic. The definitions and theorems constitute the mathematical underpinnings of the theory outlined in the main text. The conditions of Model 1 are assumed.

DEFINING RATIONALITY

Let S be the set of states of nature for a decision problem. The members of S are denoted by variables s, s', etc. and its subsets are A, B, etc. Let F be the set of all potential consequences f, g, h, etc. involved in the problem. An *act* for the problem is a function $\mathbf{f}$ from S into F.

A *choice function* is a function C mapping every set of acts for the problem to one of the set's members. In the case of sets with just two members, for any acts $\mathbf{f}, \mathbf{g}$ clearly either $C(\{\mathbf{f},\mathbf{g}\}) = \mathbf{f}$ or $C(\{\mathbf{f},\mathbf{g}\}) = \mathbf{g}$. Choice functions are interpreted behaviorally. That is, $C(\{\mathbf{f},\mathbf{g}\})$ is the act that an organism with the choice function C actually chooses or would choose if confronted with a choice between $\mathbf{f}$ and $\mathbf{g}$. For brevity, expressions of form $C(\{\mathbf{f},\mathbf{g}\}) = \mathbf{f}$ will be written $\mathbf{f} \, C \, \mathbf{g}$, which may be read "$\mathbf{f}$ is chosen over $\mathbf{g}$."

A strict preference relation $>$ between acts is definable in terms of the choice function. Clearly it should be required that if $\mathbf{f} > \mathbf{g}$ then $\mathbf{f} \, C \, \mathbf{g}$. The converse, however, is unwanted because $\mathbf{f} \, C \, \mathbf{g}$ can happen also in case of indifference. It suffices to require for $\mathbf{f} > \mathbf{g}$, beside $\mathbf{f} \, C \, \mathbf{g}$, that if the function $\mathbf{f}$ or the function $\mathbf{g}$ is modified slightly (i.e. changed on any sufficiently small set of states) then the choice between

203

them should remain unaffected. The following definition captures the idea, provided the state set is sufficiently finely partitionable.

DEFINITION A.1: For any acts $\mathbf{f}, \mathbf{g}, \mathbf{f} > \mathbf{g}$ if and only if: $\mathbf{f}$ C $\mathbf{g}$ and moreover for every consequence h in F there is a finite partition of S such that for any subset A of S that is a member of the partition,

(1) for the act $\mathbf{f}'$ such that for all s in A $\mathbf{f}'(s) = h$ and for all s in $\sim A$ $\mathbf{f}'(s) = \mathbf{f}(s)$, $\mathbf{f}'$ C $\mathbf{g}$; and

(2) for the act $\mathbf{g}'$ such that for all s in A $\mathbf{g}'(s) = h$ and for all s in $\sim A$ $\mathbf{g}'(s) = \mathbf{g}(s)$, $\mathbf{f}$ C $\mathbf{g}'$.

Indifference is defined by stipulating that $\mathbf{f} \sim \mathbf{g}$ if and only if neither $\mathbf{f} > \mathbf{g}$ nor $\mathbf{g} > \mathbf{f}$. Weak preference is defined by $\mathbf{f} \geqslant \mathbf{g}$ if and only if either $\mathbf{f} > \mathbf{g}$ or $\mathbf{f} \sim \mathbf{g}$.

The Savage postulates express various characteristics of preference that are associated with coherent rational choice. For convenience of reference they are set out below with relevant definitions interspersed. They are quoted essentially verbatim from Savage's book (1972, end papers), which may be consulted for their interpretation.

P1. The relation $\geqslant$ is a simple ordering.

D1. $\mathbf{f} \geqslant \mathbf{g}$ *given B*, if and only if $\mathbf{f}' \geqslant \mathbf{g}'$ for every $\mathbf{f}'$ and $\mathbf{g}'$ that agree with $\mathbf{f}$ and $\mathbf{g}$, respectively, on B and with each other on $\sim B$ and $\mathbf{g}' \geqslant \mathbf{f}'$ either for all such pairs or for none.

P2. For every $\mathbf{f}, \mathbf{g}$, and $B, \mathbf{f} \geqslant \mathbf{g}$ given B or $\mathbf{g} \geqslant \mathbf{f}$ given $\mathbf{B}$.

D2. $g \geqslant g'$; if and only if $\mathbf{f} \geqslant \mathbf{f}'$, when $\mathbf{f}(s) = g$, $\mathbf{f}'(s) = g'$ for every s in S.

D3. B is *null*, if and only if $\mathbf{f} \geqslant \mathbf{g}$ given B for every $\mathbf{f}, \mathbf{g}$.

P3. If $\mathbf{f}(s) = g$, $\mathbf{f}'(s) = g'$ for every s in B, and B is not null; then $\mathbf{f} \geqslant \mathbf{f}'$ given B, if and only if $g \geqslant g'$.

D4. $A \geqslant B$ if and only if $\mathbf{f}_A \geqslant \mathbf{f}_B$ or $g \geqslant g'$ for every $\mathbf{f}_A, \mathbf{f}_B, g, g'$ such that: $\mathbf{f}_A(s) = g$ for all s in A, $\mathbf{f}_A(s) = g'$ for all s in $\sim A$, $\mathbf{f}_B(s) = g$ for all s in B, $\mathbf{f}_B(s) = g'$ for all s in $\sim B$.

P4. For every $A, B, A \geqslant B$ or $B \geqslant A$.

P5. It is false that, for every $f, f', f \geqslant f'$.

P6. Suppose it is false that $\mathbf{h} \geqslant \mathbf{g}$; then, for every f, there is a (finite) partition of S such that, if $\mathbf{g}'$ agrees with $\mathbf{g}$ and $\mathbf{h}'$ agrees with $\mathbf{h}$ except on an arbitrary element of the partition, **g′** and **h′** being equal to *f* there, then it will be false that **h** ≽ **g′** or **h′** ≽ **g**.

D5. **f** ≽ *g given B* (*g* ≽ **f** given *B*); if and only if **f** ≽ **h** given *B* (**h** ≽ **f** given *B*), when **h**(*s*) = *g* for every *s*.

P7. If **f** ≽ **g**(*s*) given **B** (**g**(*s*) ≽ **f** given *B*) for every *s* in *B*, then **f** ≽ **g** given *B* (**g** ≽ **f** given *B*).

Classical decision-theoretic rationality can now be characterized in terms of these postulates.

DEFINITION A.2: ≽ is *Savage-rational* if and only if ≽ satisfies the Savage postulates P1–P7.

Observe that no mention has yet been made of environmental probabilities, fitnesses, or evolutionary stability. Savage-rationality does not involve any quantitative environmental parameters, but is rather an internal structural property of the preference relation itself. It is possible to distinguish between rationality and nonrationality in a preference relation independently of particular environmental conditions.

OPTIMALITY THEORY

In any given decision situation there is a particular probability measure *P* over the set of all subsets of *S*. Its values represent the true objective environmental probabilities of events relevant to the decision situation. *S* will be assumed here to be sufficiently finely partionable so that each event (subset of *S*) has subevents of all lesser probabilities. Also, in a particular situation each consequence *f* in *F* has an associated fitness magnitude $U(f)$ interpretable as an expected number of offspring. These fitnesses too are objectively determined. *U* will be referred to as the *fitness valuation* for the possible consequences in *F*. Mathematically *U* is a function from *F* onto some bounded set *Z* of nonnegative real numbers. *Z* is assumed to contain at least two members.

For every act **f**, *P* and *U* jointly determine an expectation $E(U(\mathbf{f}(.)))$. Biologically this expectation is the expected number of offspring for **f**, in other words the act's fitness. Here $U(\mathbf{f}(.))$ stands for the composite function that maps each *s* in *S* to the real number $U(\mathbf{f}(s))$. Statistically speaking this function is a bounded random variable on *S*. For brevity all expressions of form $E(U(\mathbf{f}(.)))$ will here be written in the shortened form $E(\mathbf{f})$.

For a choice function to be of maximal fitness it is clear there can be no acts $\mathbf{f}, \mathbf{g}$ such that $\mathbf{f} \, C \, \mathbf{g}$ and $E(\mathbf{f}) < E(\mathbf{g})$. For if there were, reversing the choice for $\mathbf{f}$ and $\mathbf{g}$ while leaving the rest of the choice function unchanged would yield a fitter choice function. That suggests the following definition of theoretical optimality:

DEFINITION A.3: C is *optimal* (with respect to P and U) if and only if for all acts $\mathbf{f}, \mathbf{g}$, if $\mathbf{f} \, C \, \mathbf{g}$ then $E(\mathbf{f}) \geq E(\mathbf{g})$.

For a decision problem that recurs over many generations it is plausible that an optimal or near-optimal choice function could evolve.

The following lemma relates preference relations to fitness expectations when the choice function is optimal.

LEMMA: If C is optimal then for all acts $\mathbf{f}, \mathbf{g}$,
   (1) $\mathbf{f} > \mathbf{g}$ if and only if $E(\mathbf{f}) > E(\mathbf{g})$;
   (2) $\mathbf{f} \sim \mathbf{g}$ if and only if $E(\mathbf{f}) = E(\mathbf{g})$;
   (3) $\mathbf{f} \geqslant \mathbf{g}$ if and only if $E(\mathbf{f}) \geq E(\mathbf{g})$.

Proof: Assume C is optimal. To prove (1) in the 'only if' direction, suppose $\mathbf{f} > \mathbf{g}$. Then by Definition A.1, $\mathbf{f} \, C \, \mathbf{g}$. Hence by Definition A3 $E(\mathbf{f}) \geq E(\mathbf{g})$. Because Z has more than one element, there must exist a consequence h such that $U(h)$ is either smaller than $E(\mathbf{f})$ or larger than $E(\mathbf{g})$. Say the latter. Then in the partition guaranteed to exist for h by Definition A.1, there must be at least one element A such that for the $\mathbf{g}'$ constructed from $\mathbf{g}$ as described in the definition, $E(\mathbf{g}') > E(\mathbf{g})$. Since $\mathbf{f} \, C \, \mathbf{g}'$ by Definition A.1, Definition A.3 gives $E(\mathbf{f}) \geq E(\mathbf{g}')$. Therefore $E(\mathbf{f}) > E(\mathbf{g})$. To prove (1) in the "if" direction, suppose $E(\mathbf{f}) > E(\mathbf{g})$. By Definition A.3, $\mathbf{f} \, C \, \mathbf{g}$. Because Z is bounded it is possible to construct a finite partition of S by dividing it up into N equiprobable subsets where N is large enough so that $(lub(Z) - glb(Z))/N < E(\mathbf{f}) - E(\mathbf{g})$. For any consequence h and any element A of this partition, with $\mathbf{f}', \mathbf{g}'$ constructed with respect to A and h as described in Definition A.1, $|E(\mathbf{f}') - E(\mathbf{f})| < E(\mathbf{f}) - E(\mathbf{g})$ and $|E(\mathbf{g}') - E(\mathbf{g})| < E(\mathbf{f}) - E(\mathbf{g})$. Hence $E(\mathbf{f}') > E(\mathbf{g})$ and $E(\mathbf{f}) > E(\mathbf{g}')$. By Definition A.3 it follows that $\mathbf{f}' \, C \, \mathbf{g}$ and $\mathbf{f} \, C \, \mathbf{g}'$, completing the requirements for $\mathbf{f} > \mathbf{g}$ under Definition A1. Parts (2) and (3) follow immediately from (1).

Optimality, where attainable, leads to rational preferences. This is shown by the following theorem.

THEOREM A.1: If C is optimal (with respect to P and U) then $\succcurlyeq$ is Savage-rational.

Proof: Assume C is optimal. Each of the Savage postulates will be derived in turn. The lemma is used ubiquitously without specific attribution.

For P1: By a 'simple ordering' Savage means a binary relation that is transitive and connected. For the proof of transitivity suppose $\mathbf{f} \succcurlyeq \mathbf{g}$ and $\mathbf{g} \succcurlyeq \mathbf{h}$. Then $E(\mathbf{f}) \geq E(\mathbf{g}) \geq E(\mathbf{h})$ and hence $\mathbf{f} \succcurlyeq \mathbf{h}$. For the connectedness recall $\succcurlyeq$ connected means that either $\mathbf{f} \succcurlyeq \mathbf{g}$ or $\mathbf{g} \succcurlyeq \mathbf{f}$. Under the definition of $\succcurlyeq$ this reduces to the statement $\mathbf{f} > \mathbf{g}$ or $\mathbf{g} > \mathbf{f}$ or neither $\mathbf{f} > \mathbf{g}$ nor $\mathbf{g} > \mathbf{f}$, which is a tautology.

For P2: It must be shown for all acts $\mathbf{f}, \mathbf{g}, \mathbf{f}', \mathbf{g}'$ that if $\mathbf{f}$ agrees with $\mathbf{g}$ and $\mathbf{f}'$ agrees with $\mathbf{g}'$ on $\sim B$ and $\mathbf{f}$ agrees with $\mathbf{f}'$ and $\mathbf{g}$ agrees with $\mathbf{g}'$ on B and $\mathbf{f} \succcurlyeq \mathbf{g}$, then $\mathbf{f}' \succcurlyeq \mathbf{g}'$ (Savage 1972, 23). Suppose $\mathbf{f} \succcurlyeq \mathbf{g}$. Then $E(\mathbf{f}) \geq E(\mathbf{g})$. But since $E(\mathbf{f}') - E(\mathbf{f}) = E(\mathbf{g}') - E(\mathbf{g})$, one has also $E(\mathbf{f}') \geq E(\mathbf{g}')$. Therefore $\mathbf{f}' \succcurlyeq \mathbf{g}'$.

For P3: Since B is not null, by D3 there exist $\mathbf{f}, \mathbf{g}$ for which $\mathbf{g} > \mathbf{f}$ given B. That is, by D1, for some $\mathbf{f}', \mathbf{g}'$ such that $\mathbf{f}'$ agrees with $\mathbf{f}$ and $\mathbf{g}'$ agrees with $\mathbf{g}$ on B and $\mathbf{f}$ agrees with $\mathbf{g}$ and $\mathbf{f}'$ with $\mathbf{g}'$ on $\sim B$, $\mathbf{g}' > \mathbf{f}'$ and consequently $E(\mathbf{g}') > E(\mathbf{f}')$. Because $\mathbf{f}'$ and $\mathbf{g}'$ differ only on B but have different expectations, B must have nonzero probability. Hence conditional probability given B is defined (Savage 1972, 264). Now consider any $\mathbf{f}, \mathbf{f}', g, g'$ such that $\mathbf{f}(s) = g$ and $\mathbf{f}'(s) = g'$ for all s in S. If $\mathbf{f} \succcurlyeq \mathbf{f}'$ given B then $E(\mathbf{f} \,|B) \geq E(\mathbf{f}' \,|B)$. Since $E(\mathbf{f} \,|B) = U(g)$ and $E(\mathbf{f}' \,|B) = U(g')$ it follows that $U(g) \geq U(g')$. Thus $g \succcurlyeq g'$. The argument is reversible.

For P4: Let $f, f', g, g', A, B, \mathbf{f}_A, \mathbf{f}_B, \mathbf{g}_A, \mathbf{g}_B$, be such that $\mathbf{f} > \mathbf{f}', \mathbf{g} > \mathbf{g}'$, $\mathbf{f}_A(s) = f$ and $\mathbf{g}_A(s) = g$ for s in A, $\mathbf{f}_A(s) = f'$ and $\mathbf{g}_A(s) = g'$ for s in $\sim A$, $\mathbf{f}_B(s) = f$ and $\mathbf{g}_B(s) = g$ for s in B, and $\mathbf{f}_B(s) = f'$ and $\mathbf{g}_B(s) = g'$ for s in $\sim B$. It has to be shown that if $\mathbf{f}_A \succcurlyeq \mathbf{f}_B$ then $\mathbf{g}_A \succcurlyeq \mathbf{g}_B$ (Savage 1972, 31). If $\mathbf{f}_A \succcurlyeq \mathbf{f}_B$ then $E(\mathbf{f}_A) \geq E(\mathbf{f}_B)$. Thus

$$P(A)U(f) + (1 - P(A))U(f') \geq P(B)U(f) + (1 - P(B))U(f').$$

Since $\mathbf{f} > \mathbf{f}'$ implies $U(f) > U(f')$ one deduces $P(A) \geq P(B)$. Following similar reasoning in the other direction for $\mathbf{g}_A, \mathbf{g}_B$ gives $E(\mathbf{g}_A) \geq E(\mathbf{g}_B)$. Hence $\mathbf{g}_A \succcurlyeq \mathbf{g}_B$.

For P5: It must be shown there exist consequences f, f' such that not $f \succcurlyeq f'$. By assumption Z has at least two members z, z' such that $z < z'$. Choose f, f' to be consequences for which $U(f) = z$ and $U(f') = z'$.

For $\mathbf{f}$, $\mathbf{f}'$ such that for all s in S, $\mathbf{f}(s) = f$ and $\mathbf{f}'(s) = f'$, $E(\mathbf{f}') > E(\mathbf{f})$ and so not $\mathbf{f} \succcurlyeq \mathbf{f}'$. By D2, not $f \succcurlyeq f'$.

For P6: Suppose $\mathbf{g} > \mathbf{h}$. Then $E(\mathbf{g}) - E(\mathbf{h}) > 0$. Partition S into N equiprobable subsets where N is large enough so that $(lub(Z) - glb(Z))$ $/ N < E(\mathbf{g}) - E(\mathbf{h})$. For any consequence f, for any element A of this partition, and for $\mathbf{g}'$ and $\mathbf{h}'$ constructed from $\mathbf{g}$, $\mathbf{h}$ with respect to A as described in P6 or Definition A.1, $E(\mathbf{g}') > E(\mathbf{h})$ and $E(\mathbf{g}) > E(\mathbf{h}')$, hence also $\mathbf{g}' > \mathbf{h}$ and $\mathbf{g} > \mathbf{h}'$. P6 follows.

For P7: Suppose $\mathbf{f} \succcurlyeq \mathbf{g}(s)$ given B for every s in B. By D1 and D5, for every $\mathbf{f}'$ agreeing with $\mathbf{f}$ on B and every $\mathbf{h}$ such that for some s in B $\mathbf{h}$ takes the value $U(g(s))$ everywhere on B and agrees with $\mathbf{f}'$ elsewhere, $\mathbf{f}' \succcurlyeq \mathbf{h}$. For all such $\mathbf{h}$, $E(\mathbf{f}') \geq E(\mathbf{h})$, so $E(\mathbf{f}')$ must equal or exceed the least upper bound of the $E(\mathbf{h})$'s. But then $E(\mathbf{f}') \geq E(\mathbf{g}')$ for $\mathbf{g}'$ agreeing with $\mathbf{g}$ on B and $\mathbf{f}$ agreeing with $\mathbf{f}'$ elsewhere. It follows that $\mathbf{f} \succcurlyeq \mathbf{g}$ given B. The proof of the parenthetical dual is similar. This concludes the proof of Theorem A.1.

RECURRENT DECISION PROBLEMS

By a *recurrent* decision problem is meant a decision situation that occurs repeatedly, confronting some or all members of the population in each generation over many successive generations. Different individuals may be presented with different acts to choose among, and the events that befall different individuals will in general differ, but the decision situation as a whole (act set, state set, consequence set, environmental probabilities, etc.) remains the same from generation to generation.

Thanks to the repetition, selective pressures have an opportunity to shape an optimal choice function via genetic modification or other inheritance mechanisms. Under favorable circumstances an innate 'lookup' capability for making optimal choices can evolve. Because suboptimal choice functions are invadable, optimality of choice is to be expected for any evolutionarily stable strategy (ESS) provided the problem recurs long enough, the available acts are recognizable, and nothing prevents natural selection from taking its course. Under such conditions we may posit:

STABILITY CONDITION 1: If C is an ESS for a recurrent decision problem then C is optimal (with respect to P and U).

This is by no means a complete definition of an ESS and no attempt will be made to formulate a complete definition here. Necessary conditions such as this one will characterize evolutionary stability sufficiently for present purposes.

Stability Condition 1 together with Theorem A.1 immediately yields

THEOREM A.2: If C is an ESS for a recurrent decision problem then $\geq$ is Savage-rational.

This completes the proof of Theorem 4.1 (main text) for the case of recurrent decision problems.

SUBJECTIVE PROBABILITIES AND UTILITIES

A standard result concerning Savage's postulates is stated below. For the proof consult Savage's book or Fishburn (1970, 192) where the theorem is conveniently consolidated.

THEOREM 4.2: If $\geq$ is Savage-rational then there exists a probability measure p over the subsets of S, and a bounded real-valued function u on the set of consequences F, such that for all acts $\mathbf{f}$, $\mathbf{g}$,

$$\mathbf{f} \geq \mathbf{g} \text{ if and only if } E(\mathbf{f}) \geq E(\mathbf{g}).$$

The measure p for which this holds is unique and the function u is unique up to a positive linear transformation.

The expectations mentioned in the theorem are understood to be calculated on the basis of p and u.

The measures p and u are interpretable as the internal subjective probability and utility measures of the Savage-rational organism. They are theoretical constructs and not necessarily directly observable, but their existence and properties are inferable from the organism's rational choice behavior.

Theorem 4.2 holds whether or not the decision problem is recurrent. In the case of a recurrent decision problem, if the choice function is an ESS then the subjective probability measure p will be identical with the objective environmental probability measure P and the subjective utility function u will match the objective fitness valuation function U up to a positive linear transformation. In this sense, for a recurrent problem the subjective probability and utility values are objectively correct.

NOVEL DECISION PROBLEMS

A decision situation that is not recurrent will be referred to as a *novel* decision problem. It may be unique to an individual or generation of individuals, or have recurred too rarely in past evolutionary time for an objectively optimal choice function to have evolved for it as a lookup instinct.

For novel decision situations there is in general no evolutionary path by which the organism can reach the theoretically optimal solution that an onlooking analyst with knowledge of P and U could arrive at. In the face of novelty the organism has no way of inheriting an innate knowledge of the objectively optimal choices, nor in general any way of deducing them from observation. A capacity to make crude probability and fitness estimates on the basis of problem-specific environmental clues could improve the decision making, but because the environmental clues are typically imperfect and incomplete (where available at all) the choices would in general still fall short of theoretical optimality.

Though consistently optimal choices may not be attainable for novel problems, in a well-adapted organism the decisions can at least be made with choice functions that are free of what might be called *structural instabilities*. A structural instability in a choice function is informally characterizable as a maladaptive combination of choices that is internally detectable and constructively repairable in a fitness-enhancing manner. To qualify as structural, the unfit choice combination must be recognizable from the structure of the choice function alone, without reference to local environmental probabilities or fitnesses. Also there must be a structurally determinate way of altering the choices in question in such a manner that the modified choice function can be expected to be fitter than the original. If not fitter in every case then it must at least tend to be fitter on average when the modification is effected over many novel problems. No general mathematical definition of structural instability will be given but examples exhibiting these properties will be encountered presently.

Choice function strategies that allow structural instabilities are invadable by strategies that avoid them. Hence, where many novel decision problems are encountered over many generations, there will be selection for a tendency to avoid structural instabilities. This imposes another necessary condition on evolutionary stability:

STABILITY CONDITION 2: If C is an ESS for a novel decision problem then there are no structural instabilities in C.

It will be seen that structural instabilities are detectable in the presence of what will be called a *reference decision problem*. A reference decision problem can be conceived as a recurrent decision problem used by the organism as an aid in solving a novel decision problem. It is helpful to think of it as a choice situation for which stability has already been reached. The innate optimal choice function for the reference problem is then available (as a 'measuring stick') to apply to the novel problem. It was seen that for a recurrent decision problem there are subjective probabilities and utilities which can be expected to approach objective accuracy as stability is approached. Once these accurate quantities are established they can be used in a manner to be explained, to detect and eliminate unstable choice combinations.

It will be assumed in reasoning about novel problems that a recurrent decision situation is available to serve as a reference problem. To simplify the reasoning it will also be assumed that this reference problem 'spans' the novel problem. By this it is meant that its consequence set contains a consequence so preferred that it is chosen over every consequence involved in the novel problem, and also a consequence so unpreferred that every consequence for the novel problem would be chosen over it. (This assumption could be weakened.)

Notice that Definition A.3 defines optimality not only with respect to the actual objective environmental P and U, but also with respect to any hypothetical probability measure p over the subsets of S and function u over F that might be of interest. The following theorem demonstrates that given stability, for any novel decision situation, hypothetical p and u will exist as subjective probability and utility measures with respect to which all the choices are optimal. There is no guarantee in this case, though, that p and u will match the objective P and U.

THEOREM A.3: If C is an ESS for a novel decision problem then there exists a probability measure p over the set of all subsets of S and a bounded real-valued function u on the set of consequences F such that C is optimal with respect to p and u.

Proof: Let S be the state set and F the consequence set for the novel decision situation. S has members s, s', etc. and subsets A, B, C, etc. F has members f, g, h, i, etc. The acts for the novel decision problem are functions **f**, **g**, etc. from S into F. The choice function among acts is C. The objective environmental probability measure is P and the objective fitness valuation function is U.

Let S_0 be the state set and F_0 the consequence set for the recurring decision situation that is to serve as the reference problem. The state set has members s_0, s_0', etc. and subsets A_0, B_0, C_0, etc. while the consequence set has members f_0, g_0, h_0, i_0, j_0, k_0, etc. The acts **f**$_0$, **g**$_0$, etc. for the recurring problem are functions from S_0 into F_0. The choice function is C_0. The objective probability measure and fitness valuation function are P_0 and U_0.

Once stability has been reached for the reference problem, its preference relation satisfies the Savage postulates by Theorem A.2. There then exist for it a probability measure p_0 and a utility function u_0 satisfying the conditions of Theorem 4.2. These subjective measures are objectively accurate reflections of P_0 and U_0. They are the reference measuring sticks known to the organism in the sense that inherited choices sufficient to determine them are available to the cognitive apparatus, which under selective pressure could come to exploit them.

The novel and the reference decision situations can be analyzed together in terms of a comprehensive *joint* decision situation in which they are both embedded. Such a joint problem can be constructed as follows. Let its consequence set F^* be $F \cup F_0$. For its state set S^* a state set is needed that is fine enough so that both S and S_0 are representable as partitionings of it. The simplest such set has as its elements all states s^* definable by conjoining the defining conditions of a member s of S with the defining conditions of a compatible state s_0 in S_0. An act for the joint decision problem is a function **f*** from S^* to F^*. The choice function among the acts is C*.

All statements concerning the novel and the reference decision situations are translatable into statements about this joint problem situation. However, to avoid cumbersome notation we will continue to use mainly the terminology introduced for the two problems taken separately, on the understanding that all statements so expressed could be reformulated as statements about the joint decision problem.

An expression of form $[f, A, g]$ refers to the act mapping all member states of A to consequence f and all member states of the complement of A to consequence g. It may be read "the gamble on A between f and g." In the special case of gambles of form $[f, A, f]$ the expression denotes a constant act that maps all states to f. In that case the specification of A is immaterial and the expression can be shortened to $[f]$. Let j_0 henceforward denote some consequence in F_0 with the special property that $[j_0]$ C^* $[f]$ for all f in F, and likewise let k_0 denote a consequence in F_0 such that $[f]$ C^* $[k_0]$ for all f in F. Such consequences exist under the assumption that the reference problem spans the novel problem.

As a first example of a structural instability suppose $p_0(B_0) > p_0(C_0)$ yet for some A, $[j_0, A, k_0]$ C^* $[j_0, B_0, k_0]$ and $[j_0, C_0, k_0]$ C^* $[j_0, A, k_0]$. Such a choice function is unstable, for it is invadable by a modified choice function $C^{*\prime}$ which is identical to C^* except that both these choices have been reversed. To see why, note that p_0 and u_0 are determinate for the organism and along with them the expectations $E([j_0, B_0, k_0])$ and $E([j_0, C_0, k_0])$ with the former determinately larger. On the other hand P is not in general determinate, so the organism has in general no way of accurately determining $E([j_0, A, k_0])$. The latter expectation is an unknown quantity X so far as data available to the organism is concerned. But whatever X may be, the adoption of C^* yields X and $E([j_0, C_0, k_0])$ for the two choices respectively, whereas $C^{*\prime}$ yields the determinately larger $E([j_0, B_0, k_0])$ and X. Hence reversing the choices is generally advantageous.

To analyze the advantage in greater depth it is necessary to take into account the occurrence probabilities of the choice pairs presented to the organism. Consider first the case in which there is equal nonzero probability P (not to be confused with P or p) of an individual being confronted with either of the above-described choice pairs. (If the same novel problem suddenly arises for many individuals in the same generation, P has a frequentist interpretation as the proportion of individuals confronted by the choice pair.) The expected advantage in fitness to be gained from changing C^* to $C^{*\prime}$ is then $P \times (E([j_0, B_0, k_0]) - X) + P \times (X - E([j_0, C_0, k_0]))$. This is a positive quantity whatever X may be, so the change is sure to improve the expectation for this decision situation. Next consider the case in which the occurrence probabilities of the two choice situations are unequal probabilities P_1 and P_2 that are unknown to the organism. The complication then arises that $P_1 \times (E([j_0, B_0, k_0]) - X) + P_2 \times (X - E([j_0, C_0, k_0]))$ is not necessarily a positive quantity. With bad luck, reversing the choices could actually lower the expectation for this particular problem situation. Nevertheless over many different novel problems the policy of always reversing such choices would be expected to yield a fitness increase on average. There is no reason to suppose the occurrence probabilities would be so systematically perverse as to block this tendency.

Suppose then that choice reversals of this kind have been made wherever applicable in the novel problem. Under a resulting choice function C* free of all instabilities of this type, for any A there must exist A_0 with the property that for all B_0 for which $p_0(B_0) < p_0(A_0)$, $[j_0, A, k_0]$ C* $[j_0, B_0, k_0]$; while for all B_0 such that $p_0(B_0) > p_0(A_0)$, $[j_0, B_0, k_0]$ C* $[j_0, A, k_0]$. Such an A_0 will be called a *commensurate reference event* for the subjective probability of A. Let p be the function for which $p(A) = p_0(A_0)$ and which similarly assigns to each other subset of S the probability under p_0 of a commensurate reference event for it in S_0.

It can be verified that the function p so constructed is a probability measure over the subsets of S which satisfies the usual (Kolmogorov) probability postulates. We indicate the proof for the postulate stating that the probability of the union of two disjoint events is equal to the sum of their separate probabilities. Suppose for two disjoint events A and B this property were to fail to obtain, say because $p(A \cup B) > p(A)$ $+ p(B)$. Then there exist A_0, B_0, and C_0 such that $p_0(A_0) = p(A) + \varepsilon$, $p_0(B_0) = p(B) + \varepsilon$, and $p_0(C_0) = p(A \cup B) - \varepsilon$ where ε is a positive number small enough so that $p_0(C_0) > p_0(A_0) + p_0(B_0)$. By construction of p, $[j_0, A_0, k_0]$ C* $[j_0, A, k_0]$, $[j_0, B_0, k_0]$ C* $[j_0, B, k_0]$, and $[j_0, A \cup B, k_0]$ C* $[j_0, C_0, k_0]$. This is a structural instability, with fitness enhanceable by reversing all three choices.

For A_0 and B_0 such that $p_0(A_0) > p_0(B_0)$, suppose for some consequence f in F, $[f]$ C* $[j_0, A_0, k_0]$ and $[j_0, B_0, k_0]$ C* $[f]$. This is a structural instability, for fitness expectations are improvable by reversing both choices. For a choice function C* free of all instabilities of this type there exist for any consequence f in F an event D_0 with the property that for all A_0 for which $p_0(A_0) > p_0(D_0)$, $[j_0, A_0, k_0]$ C* $[f]$; and for all A_0 for which $p_0(A_0) < p_0(D_0)$, $[f]$ C* $[j_0, A_0, k_0]$. The event D_0 is a *commensurate reference event* for the utility of f and may be denoted $D_0^{(f)}$. Let $u(f) = E([j_0, D_0^{(f)}, k_0]) = p_0(D_0^{(f)})u_0(j_0) + p_0(\sim D_0^{(f)})u_0(k_0)$. Repeating this construction for all consequences for the novel problem defines a utility valuation function u over F.

It remains to show that a stable choice function C must be optimal with respect to the p and u so constructed, i.e., that if $\mathbf{f}$ C $\mathbf{g}$ then $E(\mathbf{f})$ $\geq E(\mathbf{g})$ when expectation is calculated using p and u. A proof of this in full generality would involve the general definition of expectation in terms of characteristic functions (Savage 1972, Appendix I). Here we indicate only the proof for the special case in which the acts are simple gambles, though it can be generalized to gambles of arbitrary complexity. Suppose contrary to optimality that for two acts $[f, A, g]$ and $[h, B, i]$, $[f, A, g]$ C $[h, B, i]$ but $E([f, A, g]) < E([h, B, i])$. Assume for simplicity $u(f) > u(g)$, $u(h) > u(i)$, $p(A) < 1$, $p(B) > 0$. (Where these conditions are not met slight variants of the proof are called for.) Let $D_0^{(f)}$, $D_0^{(g)}$, $D_0^{(h)}$, and $D_0^{(i)}$ be commensurate reference events indicating the utilities of f, g, h, and i respectively. Then there exist A_0, B_0 for which $p_0(A_0) = p(A) + \varepsilon$, $p_0(B_0) = p(B) - \varepsilon$ where ε is small enough so that $E([[j_0, D_0^{(f)}, k_0], A_0, [j_0, D_0^{(g)}, k_0]]) < E([[j_0, D_0^{(h)}, k_0], B_0, [j_0, D_0^{(i)}, k_0]])$. One has $[[j_0, D_0^{(f)}, k_0], A_0, [j_0, D_0^{(g)}, k_0]]$ C* $[f, A, g]$, for otherwise structural instabilities could be constructed. Similarly $[h, B, i]$ C* $[[j_0, D_0^{(h)}, k_0], B_0, [j_0, D_0^{(i)}, k_0]]$. Together with the original choice these form a structural instability, with fitness improvable by reversing all three choices. This concludes the proof of Theorem A.3.

On examining Theorem A.1 it becomes clear that its proof holds for any arbitrary environmental probability measure P and fitness valuation U. Hence Theorem A.1 actually demonstrates that if there exist probabilities and fitnesses of any sort, hypothetical or not, with respect to which C is optimal, then $\succcurlyeq$ is Savage-rational. With this in mind it can be seen to be an immediate corollary of Theorem A.3 that if C is an ESS for a novel decision problem then $\succcurlyeq$ is Savage-rational. The case of recurrent problems having already been dealt with by Theorem A.2, it has at last been proved for the general case that

THEOREM 4.1: If C is an ESS then $\succcurlyeq$ is Savage-rational.

The foregoing theory need not be taken literally as a theory about temporal order of evolution – that first natural selection must solve a recurrent problem after which the solution can be used to deal with novel decision situations. Although that may be helpful as an aid to understanding, and is plausible as a historical sequence, the theorems are intended more as a demonstration that a rational capacity to solve novel problems and recurrent problems could co-evolve. It is conceivable too (though this has not been proved) that the effective equivalent of a 'virtual' reference problem having no specific application might evolve in the cognitive apparatus as a purely internal measuring stick. Were that to happen there could be Savage-rationality even in the absence of any recurrent problems, for the virtual problem would suffice.

DEDUCTIVE LOGIC

With the existence of subjective probabilities established for stable organisms, it is natural to investigate such questions as whether a very high (or low) subjective probability for one event may sometimes require a very high (low) subjective probability for some other. The question leads naturally into the study of deductive logical consequence and other deductive relationships. For the following definitions Savage-rationality is assumed, p is a variable ranging over all probability measures over the subsets of S, while $A_1, \ldots, A_N, B$ range over the subsets of S as before.

DEFINITION 5.1: B is a *logical consequence* of $A_1, \ldots, A_N$ (where $N \geq 0$) if and only if for every $\varepsilon > 0$ there exists $\delta > 0$ such that for all p, if $p(A_1), \ldots, p(A_N) > 1 - \delta$ then $p(B) > 1 - \varepsilon$.

DEFINITION 5.2: $A_1, \ldots, A_N$ (where $N > 0$) are *logically incompatible* if and only if there exists $\delta > 0$ such that there is no p for which $p(A_1), \ldots, p(A_N) > 1 - \delta$.

These definitions give rise to a theory of deductive logic consistent with the classical as outlined in the main text.

References

Adams, E. W. 1965. The Logic of Conditionals. *Inquiry* 8:166–197.

Adams, E. W. 1966. Probability and the Logic of Conditionals. In *Aspects of Inductive Logic*, J. Hintikka and P. Suppes (eds.). Amsterdam: North Holland, pp. 265–316.

Adams, E. W. 1975. *The Logic of Conditionals: An Application of Probability to Deductive Logic*. Dordrecht, Holland: D. Reidel.

Adams, E. W. 1998. *A Primer of Probability Logic*. CSLI Lecture Notes, No. 68. Stanford, Calif.: Center for the Study of Language and Information.

Allais, M. 1953. Le comportemont de l'homme rationnel devant le risque. *Econometrica* 21:503–546.

Ayer, A. J. 1956. *The Problem of Knowledge*. London: Macmillan; New York: St. Martin's Press.

Bentham, J. 1823. *An Introduction to the Principles of Morals and Legislation*. London: W. Pickering.

Black, M. 1959. *The Nature of Mathematics*. Paterson, N.J.: Littlefield, Adams & Co.

Burian, R. M. 1985. On Conceptual Change in Biology: The Case of the Gene. In *Evolution at a Crossroads: The New Biology and the New Philosophy of Science*, D. J. Depew and B. H. Weber (eds.). Cambridge, Mass.: The MIT Press, pp. 21–42.

Cabanac, M. 1992. Pleasure: The Common Currency. *Journal of Theoretical Biology* 155:173–200.

Campbell, D. T. 1974. Evolutionary Epistemology. In *The Philosophy of Karl Popper*, P. Schilpp (ed.). LaSalle, Ill.: Open Court, pp. 413–463.

Carnap, R. 1942, 1943. *Studies in Semantics*, vols. 1, 2. Cambridge, Mass.: Harvard University Press.

Carnap, R. 1950. *Logical Foundations of Probability*. Chicago: University of Chicago Press.

Carnap, R. 1952. *The Continuum of Inductive Methods*. Chicago: University of Chicago Press.

Carnap, R. 1971. A Basic System of Inductive Logic, Part 1. In *Studies in Inductive Logic and Probability*, vol. 1, R. Carnap and R. C. Jeffrey (eds.). Berkeley: University of California Press, pp. 33–166.

217

Carnap, R. 1980. A Basic System of Inductive Logic, Part 2. In *Studies in Inductive Logic and Probability*, vol. 2, R. C. Jeffrey (ed.). Berkeley: University of California Press, pp. 7–155.

Carnap, R. and R. C. Jeffrey. 1971. *Studies in Inductive Logic and Probability (I)*. Berkeley: University of California Press.

Cooper, W. S. 1978. *Foundations of Logico-linguistics: A Unified Theory of Information, Language, and Logic*. Dordrecht, Holland: D. Reidel.

Cooper, W. S. 1981. Natural Decision Theory: A General Formalism for the Analysis of Evolved Characteristics. *Journal of Theoretical Biology* 92:401–415.

Cooper, W. S. 1984. Expected Time to Extinction and the Concept of Fundamental Fitness. *Journal of Theoretical Biology* 107:603–629.

Cooper, W. S. 1987. Decision Theory as a Branch of Evolutionary Theory: A Biological Derivation of the Savage Axioms. *Psychological Review* 94:395–411.

Cooper, W. S. 1988. Is Decision Theory a Branch of Biology? In *Advances in Cognitive Science: Steps Toward Convergence*, M. Kochen and H. Hastings (eds.). Boulder, Colo.: Westview Press, pp. 7–25.

Cooper, W. S. 1989. How Evolutionary Biology Challenges the Classical Theory of Rational Choice. *Journal of Biology and Philosophy* 4:457–481.

Cooper, W. S. and R. H. Kaplan. 1982. Adaptive "Coin-flipping": A Decision-theoretic Examination of Natural Selection for Random Individual Variation. *Journal of Theoretical Biology* 94:135–151.

Crow, J. F. and M. Kimura. 1970. *An Introduction to Population Genetics Theory*. New York: Harper and Row.

Darwin, Charles. 1871. *The Descent of Man*. London: J. Murray.

Dawkins, R. 1982. *The Extended Phenotype: The Gene as the Unit of Selection*. Oxford: W. H. Freeman & Co.

Edwards, W. 1954. The Theory of Decision Making. *Psychological Bulletin* 51:380–417.

Edwards, W. 1961. Behavioral Decision Theory. *Annual Review of Psychology* 12:473–498.

Eells, E. and B. Skyrms. 1994. *Probability and Conditionals*. Cambridge: Cambridge University Press.

Ewing, A. C. 1940. The Linguistic Theory of *a priori* Propositions. *Proceedings of the Aristotelian Society* 40:207–244.

Fischhoff, B., B. Goitein, and Z. Shapira. 1982. The Experienced Utility of Expected Utility Approaches. In *Expectations and Actions*, N. Feather (ed.). Hillsdale, N.J.: Erlbaum Associates, pp. 117–141.

Fishburn, P. C. 1970. *Utility Theory for Decision Making*. Publications in Operations Research No. 18. New York: John Wiley & Sons.

Fishburn, P. C. 1981. Subjective Expected Utility: A Review of Normative Theories. *Theory and Decision* 13:139–199.

Fisher, I. 1918. Is "Utility" the Most Suitable Term for the Concept It Is Used To Denote? *American Economic Review* 8:335–337.

Futuyma, D. J. 1986. *Evolutionary Biology* (2nd ed.). Sunderland, Mass.: Sinauer Associates.

References

Gillespie, J. H. 1974. Natural Selection for Within-generation Variance in Offspring Number. *Genetics* 76:601–606.

Gillespie, J. H. 1977. Natural Selection for Variances in Offspring Numbers: A New Evolutionary Principle. *American Naturalist* 111:1010–1014.

Godfrey-Smith, P. 1996. *Complexity and the Function of Mind in Nature.* Cambridge, England: Cambridge University Press.

Goodman, N. 1955. *Fact, Fiction and Forecast.* Cambridge, Mass.: Harvard University Press.

Gould, S. J. and R. C. Lewontin. 1979. The Spandrels of San Marco and the Panglossian Paradigm: A Critique of the Adaptationist Programme. *Proceedings of the Royal Society* B205:581–598.

Gould, S. J. and E. S. Vrba. 1982. Exaption: A Missing Term in the Science of Form. *Paleobiology* 8:4–15.

Hartl, D. L. 1980. *Principles of Population Genetics.* Sunderland, Mass.: Sinauer Associates.

Heyting, A. 1964. Disputation. In *Philosophy of Mathematics: Selected Readings*, P. Benacerraf and H. Putnam (eds.). Englewood Cliffs, N.J.: Prentice-Hall.

Jeffrey, R. C. 1964. If (abstract). *Journal of Philosophy* 61:702–703.

Jeffrey, R. C. 1983. *The Logic of Decision* (2nd ed.). Chicago: University of Chicago Press.

Kahneman, D., P. Slovic, and A. Tversky (eds.). 1982. *Judgment under Uncertainty: Heuristics and Biases.* Cambridge, England: Cambridge University Press.

Kaplan, R. H. and W. S. Cooper. 1984. The Evolution of Developmental Plasticity in Reproductive Characteristics: An Application of the "Adaptive Coin-flipping" Principle. *American Naturalist* 123:393–410.

Kline, M. 1980. *Mathematics: The Loss of Certainty.* Oxford: Oxford University Press.

Kuhn, T. S. 1957. *The Copernican Revolution: Planetary Astronomy in the Development of Western Thought.* Cambridge, Mass.: Harvard University Press.

Levins, R. 1962. Theory of Fitness in a Heterogeneous Environment. *American Naturalist* 96:361.

Lewis, D. 1981. Causal Decision Theory. *Australasian Journal of Philosophy* 59:5–30.

Lewontin, R. C. 1961. Evolution and the Theory of Games. *Journal of Theoretical Biology* 1:382–403.

Lomnicki, A. 1988. *Population Ecology of Individuals.* Princeton: Princeton University Press.

Lorenz, K. 1941. Kant's Doctrine of the *A Priori* in the Light of Contemporary Biology. Translated in *Learning, Development and Culture: Essays in Evolutionary Epistemology*, H. C. Plotkin (ed.). New York: Wiley (1982), pp. 121–143.

Luce, R. D. and P. Suppes. 1965. Preference, Utility, and Subjective Probability. In *Handbook of Mathematical Psychology*, vol. 3, R. D. Luce, R. R. Bush, and E. Galanter (eds.). New York: John Wiley & Sons, pp. 249–410.

MacArthur, R. H. and E. O. Wilson. 1967. *The Theory of Island Biogeography.* Princeton, N.J.: Princeton University Press.

Mates, B. 1972. *Elementary Logic* (2nd ed.). New York: Oxford University Press.

May, R. M. 1973. On Relationships among Various Types of Population Models. *The American Naturalist* 107:46–57.

Maynard Smith, J. 1978. Optimization Theory in Evolution. *Annual Review of Ecology and Systematics* 9:31–56.

Maynard Smith, J. 1982. *Evolution and the Theory of Games.* Cambridge, England: Cambridge University Press.

Maynard Smith, J. 1984. Game Theory and the Evolution of Behavior. *The Behavioral and Brain Sciences* 7:95–125.

Maynard Smith, J. and G. R. Price. 1973. The Logic of Animal Conflict. *Nature* 246:15–18.

McGee, V. 1994. Learning the Impossible. In *Probability and Conditionals*, E. Eells and B. Skyrms (eds.). Cambridge: Cambridge University Press, pp. 179–199.

Medawar, P. B. and J. S. Medawar. 1977. *The Life Science: Current Ideas of Biology.* New York: Harper and Row.

Mertz, D. B. 1970. Notes on the Methods Used in Life History Studies. In *Readings in Ecology and Ecological Genetics*, J. H. Connell, D. B. Mertz, and W. W. Murdoch (eds.). New York: Harper and Row, pp. 4–17.

Mills, S. K. and J. H. Beatty. 1979. The Propensity Interpretation of Fitness. *Philosophy of Science* 46:263–288.

Nagel, E. 1961. *The Structure of Science.* New York: Harcourt, Brace and World.

Needham, J. 1954. *Science and Civilization in China*, vol. 2. Cambridge: Cambridge University Press.

Oppenheim, P. and H. Putnam. 1958. Unity of Science as a Working Hypothesis. In *Concepts, Theories, and the Mind-Body Problem*, Minnesota Studies in the Philosophy of Science, vol. II, H. Feigl, M. Scriven, and G. Maxwell (eds.). Minneapolis: University of Minnesota Press, pp. 3–36.

Pielou, E. C. 1977. *Mathematical Ecology.* New York: John Wiley & Sons.

Polya, G. 1954. *Mathematics and Plausible Reasoning*, vol. 2: *Patterns of Plausible Inference.* Princeton, N.J.: Princeton University Press.

Popper, K. 1963. *Conjectures and Refutations.* London: Routledge & Kegan Paul; New York: Basic Books.

Putnam, H. 1969. Is Logic Empirical? In *Boston Studies in the Philosophy of Science*, vol. v., R. S. Cohen and M. R. Wartofsky (eds.). Dordrecht, Holland: D. Reidel, pp. 216–241.

Putnam, H. 1975. *Mathematics, Matter and Method.* Philosophical Papers, vol. I, 2nd. ed. Cambridge: Cambridge University Press.

Quine, W. V. O. 1958. *Mathematical Logic* (rev. ed.). Cambridge, Mass.: Harvard University Press.

Raiffa, H. 1968. *Decision Analysis: Introductory Lectures on Choices Under Uncertainty.* Reading, Mass.: Addison-Wesley.

Ramsey, F. P. 1931. *The Foundations of Mathematics and Other Logical Essays.* New York: Harcourt Brace.

Real, L. A. 1980. Fitness, Uncertainty, and the Role of Diversification in Evolution and Behavior. *American Naturalist* 115:623.

Reichert, S. E. and P. Hammerstein. 1983. Game Theory in the Ecological Context. *Annual Review of Ecology and Systematics* 14:377–409.

Resnik, M. D. 1980. *Frege and the Philosophy of Mathematics*. Ithaca, N.Y.: Cornell University Press.

Robson, A. J. 1996. A Biological Basis for Expected and Non-expected Utility. *Journal of Economic Theory* 68:397–424.

Ruse, M. 1986a. *Taking Darwin Seriously: A Naturalistic Approach to Philosophy*. Oxford: Blackwell.

Ruse, M. 1986b. Intelligence and Natural Selection. In *Intelligence and Evolutionary Biology*, H. J. Jerison and I. Jerison (eds.). NATO Advanced Science Institutes Series, Series G: Ecological Sciences 17:13–34. New York: Springer Verlag.

Ruse, M. 1989. The View from Somewhere: A Critical Defense of Evolutionary Epistemology. In *Issues in Evolutionary Epistemology*, K. Hahlweg and C. A. Hooker (eds.). Albany, N.Y.: State University of New York Press, pp. 185–228.

Russell, B. 1919. *Introduction to Mathematical Philosophy*. London: G. Allen & Unwin.

Savage, L. 1972. *The Foundations of Statistics*. (2nd ed.). New York: Dover.

Schaffner, K. F. 1977. Reduction, Reductionism, Values, and Progress in the Biomedical Sciences. In *Logic, Laws, and Life*, R. G. Colodny (ed.). Pittsburgh: University of Pittsburgh Press, pp. 143–171.

Schaffner, K. F. 1993. *Discovery and Explanation in Biology and Medicine*. Chicago: University of Chicago Press.

Skyrms, B. 1984. *Pragmatics and Empiricism*. New Haven: Yale University Press.

Skyrms, B. 1987. Dynamic Coherence and Probability Kinematics. *Philosophy of Science* 54:1–20.

Skyrms, B. 1990. *The Dynamics of Rational Deliberation*. Cambridge, Mass.: Harvard University Press.

Skyrms, B. 1994. Darwin Meets the Logic of Decision: Correlation in Evolutionary Game Theory. *Philosophy of Science* 61:503–528.

Skyrms, B. 1996. *Evolution of the Social Contract*. Cambridge: Cambridge University Press.

Skyrms, B. 1997. Game Theory, Rationality and Evolution. In *Structures and Norms in Science*, M. L. Dalla Chiara, K. Doets, D. Mundici, J. van Benthem (eds.). Netherlands: Kluwer Academic Publishers, pp. 73–86.

Sober, E. 1981. The Evolution of Rationality. *Synthese* 46:95–120.

Sober, E. 1998. Six Sayings about Adaptationism. In *The Philosophy of Biology*, D. L. Hull and M. Ruse (eds.). Oxford: Oxford University Press, pp. 72–86.

Sober, E. and D. S. Wilson. 1994. A Critical Review of Philosophical Work on the Units of Selection Problem. *Philosophy of Science* 61:534–555.

Stearns, S. C. 1976. Life-History Tactics: A Review of the Ideas. *Quarterly Review of Biology* 51:3–47.

Stearns, S. C. 1989. The Evolutionary Significance of Phenotypic Plasticity. *Bioscience* 39:198–207.

Stearns, S. C. 1992. *The Evolution of Life Histories*. New York: Oxford University Press.

Tarski, A. 1956. *Logic, Semantics, Metamathematics*. London: Oxford University Press.

Templeton, A. R. and E. D. Rothman. 1974. Evolution in Heterogeneous Environments. *American Naturalist* 108:409–428.

Tversky, A. and D. Kahneman. 1992. Advances in Prospect Theory: Cumulative Representation of Uncertainty. *Journal of Risk and Uncertainty* 5:297–323.

Vollmer, G. 1987. On Supposed Circularities in an Empirically Oriented Epistemology. In *Evolutionary Epistemology, Rationality, and the Sociology of Knowledge*, G. Radnitzky and W. W. Bartley III (eds.). La Salle, Ill.: Open Court, pp. 163–200.

von Neumann, J. and O. Morgenstern. 1953. *Theory of Games and Economic Behavior* (3rd ed.). Princeton, N.J.: Princeton University Press.

Waddington, C. H. 1957. *The Strategy of the Genes.* London: George Allen and Unwin.

Whitehead, A. N. and B. Russell. 1910. *Principia Mathematica*. Cambridge: Cambridge University Press.

Wiegert, R. G. 1974. Competition: A Theory Based on Realistic, General Equations of Population Growth. *Science* 185:539–542.

Williams, G. C. 1985. A Defense of Reductionism in Evolutionary Biology. *Oxford Surveys in Evolutionary Biology* 2:1–17.

Wilson, E. O. 1978. *On Human Nature*. Cambridge, Mass.: Harvard University Press.

Index

evolutionary biology, 2, 8, 21
evolutionary epistemology, 16, 123, 200
evolutionarily stable strategy (ESS), 76
Ewing, A. C., 178
expected value, 51–3
expected time to extinction (ETE), 166–8

Fischhoff, B., 158
Fishburn, P. C., 74, 75, 209
Fisher, I., 63
fitness, 37–40, 48, 52–3, 60–1, 168–9
formalism, 194–5
Frege, G., 15, 126, 133, 196
Futuyma, D. J., 143

game theory, 14, 21, 153
geometric mean, 150–1, 158, 165, 168
Gillespie, J. H., 150
Gödel, Kurt, 106, 127, 196
Godfrey-Smith, P., 58, 146
Goodman, N., 121
Gould, S. J., 40, 41
grue, 121

Hammerstein, P., 14
Hartl, D. L., 52
Heyting, 195
Hilbert, David, 196
Hume, David, 117, 120, 173
hypothetical syllogism, 98, 112–13

if-then, 97, 99, 112–14, 199
incompleteness theorem, 106, 127
intransitivity, 78–80
intuitionism, 196

Jeffrey, R. C., 15, 63, 75, 90, 112, 115

Kahneman, D. P., 171
Kant, E., 16
Kaplan, R., 146, 154
Kimura, M., 23, 150
Kline, M., 178
Kronecker, 196
Kuhn, T. S., 2

Leibniz, 178
Levins, R., 153
Lewis, D., 81
Lewontin, R. C., 14, 41, 62
linguistics, 199
logic
  classical, 2, 6–7, 188
  deductive, 6, 20–1, 90–108, 183–4
  evolutionarily stable, 132, 158, 164, 166
  higher order, 106–7
  inductive, 6, 20–1, 69–89, 109–24
  laws of, 3–5, 19
  nonclassical, 146–72, 185–7, 200
  nonstandard, *see* nonclassical
  predicate, 103–6
  prescriptive, 107–8
  propositional, 94–103
logical certainty, 180
logical consequence, 94–6, 101, 102, 216
logical empiricism, 193
logical incompatibility, 96–7, 101, 102, 216
logicism, 15, 125–9
logistic growth, 137–41, 145